# STUDIES IN INDUCTIVE PROBABILITY AND RATIONAL EXPECTATION

# SYNTHESE LIBRARY

MONOGRAPHS ON EPISTEMOLOGY,

LOGIC, METHODOLOGY, PHILOSOPHY OF SCIENCE,

SOCIOLOGY OF SCIENCE AND OF KNOWLEDGE,

AND ON THE MATHEMATICAL METHODS OF

SOCIAL AND BEHAVIOURAL SCIENCES

*Managing Editor:*

JAAKKO HINTIKKA, *Academy of Finland and Stanford University*

*Editors:*

ROBERT S. COHEN, *Boston University*

DONALD DAVIDSON, *University of Chicago*

GABRIËL NUCHELMANS, *University of Leyden*

WESLEY C. SALMON, *University of Arizona*

VOLUME 123

# STUDIES IN INDUCTIVE PROBABILITY AND RATIONAL EXPECTATION

*by*

THEO A. F. KUIPERS

*Department of Philosophy, University of Groningen*

SPRINGER-SCIENCE+BUSINESS MEDIA, B.V.

Library of Congress Cataloging in Publication Data

Kuipers, Theodorus A. F. 1947–
   Studies in inductive probability and rational expectation.

   (Synthese library; v. 123)
   Bibliography: p.
   Includes indexes.
   1.  Probabilities.   2.   Induction (Mathematics).
   3.   Knowledge, Theory of.   4.   Analysis (Philosophy).
I.   Title.
QA273.K796            519.2            78–677

ISBN 978-94-009-9832-2            ISBN 978-94-009-9830-8 (eBook)
DOI 10.1007/978-94-009-9830-8

*To Inge*

# ACKNOWLEDGEMENTS

During the research for this book and its completion I have been greatly indebted to Prof. J. J. A. Mooij and Prof. A. J. Stam. The personal encouragement and the philosophical criticism by Prof. Mooij combined with the criticism by and great help on mathematical aspects from Prof. Stam constituted the best conditions one could wish for individual, interdisciplinary research in the common domain of philosophy and mathematics. Without them this book would never have been written.

I would also like to thank Prof. E. M. Barth, Dr. J. F. A. K. van Benthem, Dr. G. Berger, Dr. R. Cooke, Prof. W. K. Essler, Prof. R. C. Jeffrey, Prof. I. Niiniluoto and M. E. de Zoeten for their criticism and suggestions at particular stages of the research.

The main point of departure for the research has been the relevant work by Prof. R. Carnap and Prof. J. Hintikka. Without their writings this book could not have been written. From a certain stage on, a new system, invented by Prof. J. Hintikka and Prof. I. Niiniluoto, played a decisive role in the research. The publications by Prof. W. Stegmüller, Prof. R. Hilpinen and Dr. J. Pietarinen have also been a great help to me.

Finally I am very grateful to Prof. Hintikka and Prof. Niiniluoto for their recommendation to have the book published as volume 123 in the Synthese Library.

*Department of Philosophy*                                   *August, 1977*
*University of Groningen*
*The Netherlands*

# TABLE OF CONTENTS

# CHAPTER 1

# INTRODUCTION

## 1. CONCEPT EXPLICATION

One of the most important developments in philosophy during the last hundred years seems to be a methodological one. More than before, philosophers are aware of the fact that the process of tackling philosophical problems has to start with the clarification (or analysis) of the concepts involved. It is certainly true that there have been, in the history of Western philosophy and science, considerable efforts to analyse concepts and the formulation of problems, and historical research shows in many cases that these efforts have been of great importance to eventual achievements. So, the practice of clarification is not entirely new. What is rather new is the explicit recognition by many philosophers that clarification is an indispensable step in solving problems. Unfortunately, this does not mean that all these philosophers actually practise conceptual analysis. On the contrary, many of them go on to treat problems stated in vague terms and they accept, or do not realize, that their conclusions are at least as vague as the concepts involved.

It is not true either that the practitioners of conceptual analysis do form a homogeneous group. At least three fundamentally different theoretical positions can be distinguished:

*the essentialistic position* (Husserl): many important concepts may seem at first sight rather vague but yet they have a clear meaning, although it is very difficult to grasp that meaning;

*the adaptivistic position* (Wittgenstein, Ryle and Austin): concepts usually have precise rules of usage, and therefore a clear 'meaning in use', although it is not easy to make the rules explicit; and

*the constructivistic position* (Russell and Carnap): most concepts do not have a clear meaning, but they can get one (or more) if our attempts at creating such a clear meaning are successful.

No doubt, in the last fifty years, non-essentialistic conceptual analysis is being practised most frequently and most thoroughly, but not exclusively, in what is called analytic philosophy. Though the name might

1

suggest it, it is not correct to circumscribe analytic philosophy by definition as 'philosophy restricted to (non-essentialistic) conceptual analysis', for that restriction suggests an exclusive interest in problems as far as they are conceptual. More and more, analytic philosophers have become interested in substantial problems and it is primarily their *attitude* to problems which differentiates them from other philosophers, i.e., their emphasis on conceptual analysis based on the conviction that analysis is the key to the ultimate solution of conceptual and substantial problems.

We do not want to conceal the fact that many famous analytic philosophers have been of the opinion that philosophical problems will not be solved in a strict sense but that they will disappear by proper conceptual analysis. However, in the last twenty years the number of defenders of this position has certainly decreased. This is strongly related to the decay of 'ordinary language philosophy', that brand of analytic philosophy which was exclusively based on the adaptivistic position in conceptual analysis.

On the other hand, constructivistic conceptual analysis, for which we shall also use Carnap's favourite term 'concept explication', has become, more and more, important in analytic philosophy. Many activities in so-called philosophical logic belong in this view to analytic philosophy. It is quite imaginable that in philosophy a process will take place which has many similarities with a development in the exact sciences during the last two centuries. In the heydays of classical science, physics and mathematics were strongly interwoven. But gradually mathematics grew out to a separate discipline (as it was in ancient times) and, as a bridge with physics, there arose (all kinds of) applied mathematics. Analogously, it is conceivable that there will be in the future, besides 'substantial philosophy', the separate discipline of meta-philosophy and, to fill the gap, something like applied metaphilosophy. Modal logic and its (philosophical) applications provide cases in point.

The present view on analytic and meta-philosophy raises the question 'What makes a problem philosophical?' and the question 'What makes a non-conceptual problem substantial?' In our opinion the first question is not really important. Answers need to be such that the classification of a problem as a philosophical one does not prevent us from relevant empirical investigations. The question as to the substantiveness of problems may be intriguing but, at the same time, it seems impossible to give a satisfactory general answer. This does not exclude the possibility of such a qualification for solved problems. In other words the selection of problems for research in philosophy, and therefore in metaphilosophy, cannot be based on rules that avoid spending time on pseudo-problems. Of course, this implies that, if one succeeds in demonstrating convincingly the pseudo-character of a problem by giving its 'solution', the time spent on it need not be seen as wasted.

We conclude this section with a brief statement of the criteria for concept explication as they have been formulated in several places by Carnap, Hempel and Stegmüller. Hempel's account ([13], Chapter 1) is still very adequate for a detailed introduction.

The process of explication starts with the identification of one or more vague and, perhaps, ambiguous concepts, the so-called explicanda. Next, one tries to disentangle the ambiguities. This, however, need not be possible at once. Ultimately the explicanda are to be replaced (not necessarily one by one) by certain counterparts, the so-called explicata, which have to conform to four requirements. They have to be as precise as possible and as simple as possible. In addition, they have to be useful in the sense that they give rise to the formulation of theories and the solution of problems. The three requirements of preciseness, simplicity and usefulness, have of course to be pursued in all concept formation.

The following and last requirement is a characteristic of concept explication: the explicata have to be (perhaps in a complex way) similar to the explicanda, but it is impossible to give a general criterion for similarity; this has to be judged in concrete cases. For that purpose, one usually works with so-called conditions of adequacy. Such conditions are considered to be fundamental aspects of the explicanda and for that reason they are imposed on the explicata as necessary conditions.

It may well be that the formulation of conditions of adequacy leads to the discovery of ambiguities in the explicanda, in particular if the proposed conditions of adequacy turn out to be incompatible. In such a case we may come to the conclusion that there are two different concepts involved or, alternatively, that one of the conditions has to be rejected on second thoughts.

The final explicata may of course have properties which we did not recognize in the explicanda; this will even occur frequently and for that reason concept explication is a form of creation of meaning. This brings us to an important related aspect. In many cases we are inclined to talk about intuitions concerning the explicanda, on the basis of which the conditions of adequacy are introduced. Consequently, if explicata become accepted the created meanings indicated above turn into created or acquired intuitions and if conditions of adequacy turn out to be incompatible we may conclude that we have to suppress a certain antecedent intuition, provided that we do not see good reasons for distinguishing different meanings.

## 2. OBJECTIVES AND SURVEY

One of the best known explication programs has been set up by Carnap around the concepts of confirmation and rational belief under the heading of inductive logic. The verificationist standpoint of the heydays of the Vienna Circle with respect to the meaningfulness of statements in general and theories in particular had to be given up in favour of a 'confirmationist' point of view: theories were no longer required to be verifiable (by finite evidence) but confirmable and this weakening asked for an explication of the notion of degree of confirmation.

Carnap [2] formulated the project extensively, but later on the proposed probabilistic explicata turned out to be more appropriate for the notion of rational degree of belief (see, e.g., [4]). But Carnap went on to speak also of the notion of logical probability, which he saw as a concept of probability opposed to the statistical concept used in applied probability theory.

According to Carnap, logical probability had to be defined on the sentences of a language. Later on he preferred a model-theoretical approach. The former approach can be found in Carnap [2, 3] and also in Carnap and Stegmüller [8], the latter in Carnap [7] and in Stegmüller [34]. Both approaches may be called logical (or even linguistic), for the model-theoretical approach is also based on the sentences of a (first-order) language.

We have always felt uneasy with these approaches and our *first objective* has been to reformulate the entire program in set-theoretical terms, which is nowadays the standard approach in (mathematical) probability theory. The primary motivation for such a reformulation was not so much that we expected real progress in the explication program but that we hoped to get a better view on that program.

This hope was based on the intuitive feeling that the concrete results obtained thus far could be restated in the standard way. In our opinion there were two concrete results worth studying: Carnap's continuum of inductive methods (or $\lambda$-continuum) published in 1952 [3] and

Hintikka's two-dimensional continuum of inductive methods (or $\alpha$–$\lambda$-continuum) published in 1966 [16]. That Carnap's system, and even its generalization by Stegmüller [34], could be reformulated in standard terms was clear to us from the beginning. Chapter 5 is devoted to the resulting C- and GC-systems. However, it seemed rather difficult to reconstruct Hintikka's system in the same straightforward way, although it was strongly related to Carnap's continuum.

Fortunately, it turned out, later on, that a class of systems, containing the $\alpha$–$\lambda$-systems as special cases, could be reformulated rigorously. Chapter 6 treats these systems. They have been called Hintikka-systems (H-systems) because they constitute the background system for the $\alpha$–$\lambda$-system (for which the axiomatization, as a special case of H-systems, is completed in Chapter 7, Section 5).

The main inspiration for that reformulation came from a new system presented by Hintikka and Niiniluoto in 1974 [19] and our intuitive feeling that that system was not as new as it seemed at first sight. Such a (NH-) system turned out to be equivalent to a special type of H-system (an SH-system). Compared with the axiomatization of C-systems, NH-systems seem the simplest generalization of C-systems, whereas the SH-axiomatization makes explicit that these assign, in appropriate applications, non-zero probabilities to universal statements. The combination of both aspects suggested that we call these systems Universalized Carnapian (UC-) systems. In our opinion the equivalence theorem between NH- and SH-systems is the most important concrete result of this study; it will also be published separately in an issue of *Synthese* [26].

The *second objective* of our research was to give a rigorous definition of the contexts in which the C- and H-systems can be applied such that the qualification as rational (system of) degrees of belief (and expectation) is really justified For, according to our opinion, Carnap's expositions in this respect are extremely vague, but it may well be that this is, to a large degree, due to his logical approach. On the other hand, the expositions of Hintikka and his students Hilpinen [15] and Pietarinen [29] are mostly clear on this point, not in the least, perhaps, because they do not bother so much about the technical formulation but concentrate on the concrete problem of assigning probabilities to singular and universal (interpreted) statements about an infinite universe. But, from the standard point of view that specific context of application turns up as a special case of a more abstract situation. Chapter 7 contains the general characterization of the contexts in which the application of C- and UC-systems seems adequate: the closed and open multinomial contexts respectively.

In Chapter 7 we also try to argue that Carnap's intended application for C-systems has to be construed as random sampling without replacement in a finite universe or, more generally, as a hypergeometric context, but also that this application is highly objectionable. Our original idea was that in such a context so-called negative inductive GC-systems were appropriate, and this explains why we pay so much attention to them in Chapter 5.

So far we have sketched the results of the reformulation of Carnap's program as far as C- and H-systems are concerned. But a more general reconstruction was also intended and that is given in Chapter 4. There we define a paradigmatic context in terms of theories and (evidence obtained from) experiments. This definition was originally suggested by Hintikka's formulation of $\alpha$–$\lambda$-systems and it was developed in agreement with the context underlying the original point of departure of Carnap in philosophy of science, i.e., the confirmation of theories in the light of evidence. On the basis of such a paradigmatic context we define then a so-called formally rational expectation pattern. It seemed more adequate to speak only of degrees of belief in case of theories. We speak of degrees of prediction in relation to (evidence produced by) experiments and of degrees of expectation in case of combinations of theories and evidence, so-called perspectives. In this terminology degrees of belief and degrees of prediction are degrees of expectation related to perspectives of special kinds.

Chapter 2 contains a general outline of our view on the philosophical significance of the program. In Chapter 3 we present the main logical and mathematical definitions and theorems that are needed. The section on probability theory need not be worked through directly; it can be consulted if necessary. Chapter 8 contains three sections in which some technical problems are formulated. The last three sections are devoted to related topics. In the final chapter (Chapter 9) we shall make some concluding remarks about the prospects for the reconstructed explication program.

The reconstructive character of our research brings about that many results, especially in the chapters on GC- and H-systems, are not new but (generalized) reformulations of results that have been obtained by Carnap, Hintikka and others. But there are also a number of sections in which new information about these systems is given. Moreover, the objective interpretations (or models) for these systems (Chapter 5, Sections 3.3 and 4.3; Chapter 6, Section 5) are, as far as we know, entirely new. The thesis, related to the model in Chapter 5, Section 3.3, that GC-systems, and therefore C-systems, are, technically speaking, equivalent to (generalized) Polya-systems, described by Polya in 1931 [30], should also be mentioned here. A direct extension of H-systems to infinitely many predicates is to be found in Chapter 6, Section 10. The generalization of H-systems, analogous to that of C-systems to GC-systems, is contained in Chapter 6, Section 11. Chapter 6 concludes with a survey of systems in Section 12.

In Section 5 of Chapter 7 we pay attention to the constitutional distributions proposed by Hintikka (the $\alpha$-distribution, on which the $\alpha$–$\lambda$-continuum is based) and the one proposed by ourselves [25]. Special attention is asked for the special case of the latter, suggested by Carnap. The resulting *UC*-system, about which Section 3 of Chapter 8 contains an intriguing observation of its limit behaviour, is in our opinion the most sophisticated application of the classical principle of indifference.

The main philosophical thesis is contained in Section 8 of Chapter 6. There we argue, in our opinion conclusively, on the basis of the above indicated equivalence theorem, that a general (intuitive) resistance to assign non-zero probabilities to universal hypotheses is misplaced.

This survey may give a chaotic impression of the order of chapters and sections. However, the reader will observe that the order of chapters is determined by degrees of generality and the distinction between pure systems and their philosophical applications. The latter distinction is a direct consequence of our approach and it may be seen as *the main methodological change* with respect to Carnap's program.

# CHAPTER 2

# COGNITIVE RATIONALITY

## 1. ON THE EXPLICATION OF THE CONCEPT OF RATIONALITY

Rationality is a much-disputed value in modern (Western) science and society. There are at least two reasons for its controversial status. First, there is no agreement about the precise meaning of the term: the concept is rather vague. Second, there is no agreement in what sense (meaning) it is a value worth being pursued. Here we are primarily interested in the meaning of the concept. In this section we shall make some general remarks about a program for its explication. This program has been sketched in J. Mooij's inaugural address *Aspecten van redelijkheid* (*Aspects of Rationality* [28]) in which, generally speaking, Carnap's program for the explication of cognitive rationality has been extended to a program including the rationality of preferences ('desiderative' rationality) and the rationality of decisions (practical rationality).

In everyday and scientific discourse we talk about rational expectations, beliefs, predictions, preferences, decisions, actions, rules, values, principles, procedures, methods, reconstructions, discussions and human beings. On the object-level the program is restricted to rational expectations, preferences and decisions. This restriction is, however, in no sense fundamental. On the meta-level the program needs the introduction of principles, rules and values which are or can be considered to be rational with respect to expectations, preferences and decisions. It is important to note that such maxims need not be defensible as universally rational. In other words their rationality may be claimed only with respect to particular types of problems or situations or, as we call them, types of contexts. Changing the context, which is supposed to include widening and narrowing the context, may let us set aside a maxim, also when it can still be applied. As a consequence, calling expectations, preferences and decisions rational is always related to a particular type of context, too.

The foregoing does not exclude, however, that some maxims are considered to be universally rational. For example, in order to be rational, real-valued expectation within a well-defined context needs to be probabilistic, and preferences within a fixed context need to be transitive.

Such maxims may be said to be formally rational in contradistinction to contextually, or materially, rational maxims, such as, for instance, the rule to make no differences in expectations or preferences if you are unable to give defensible reasons for such differences. We do not claim that the difference between formally and materially rational maxims can be based on a structural difference within the set of all possible maxims, but this seems a problem which is worth being studied.

The restriction to expectations, preferences and decisions is a provisional one and one might even argue that the explication of their rationality will probably be of considerable importance for the explication of, for instance, rational procedures and discussions. In most cases the role of the former explications in the latter will be very complex, in other cases it may be rather easy. For example, one might argue that a sufficient condition for being a rational action, within a particular context, is that it is, or can be, based on a rational decision. Similarly, rational decisions can be based on rational expectations and rational preferences in combination with a decision rule which is considered to be rational with respect to that particular type of context (e.g., the rule of maximizing expected utility).

One might think that the program leads, ideally, to unique expectations, preferences and decisions in a fixed context. Our position in this respect is two-fold. First, there are certainly contexts in which this happens, e.g., in the case where we know the objective probabilities of an experimental process, rational expectations need, in our opinion, to coincide with these probabilities. Second, as long as we are intuitively inclined to say that there should be different ways to be rational, our explication needs to leave room for such differences. Here we have an important difference between the notions of truth and rationality. If two theories are incompatible, at most one of them can be true. On the other hand, conflicting expectation or preference patterns may both be rational. The same holds for conflicting maxims except in the case where one of them is considered to be formally, and therefore universally, rational. In other words the program is not directed to the elimination of subjective differences as such, but to proposals for defensible maxims.

This aspect of the program is strongly related to a general objective of concept explication: the development of intuitions and conventions. The limits for this activity are determined by the intuitions and, perhaps also, by conventions we do not want to give up.

One of the main discussions with respect to the question, mentioned at the beginning of this section, in what sense the value of rationality is worth being pursued, can be stated in terms of a historical thesis. In the course of (Western) history the concept of rationality has gradually been restricted to cognitive rationality and to what is called means–end or instrumental rationality, based on cognitive rationality. But it is argued that instrumental rationality is only worth being pursued if it can be embedded in a more enclosing meaning of the term. It is not altogether clear whether this plea for reconsideration of the value is a direct plea for substantial values (e.g., progress or emancipation) considered to be rational, or an indirect plea for such values by suggesting rational rules of procedure, e.g., in ethical matters. The latter possibility might be based on the expectation that (truly) rational procedures work in favour of (consensus about) the values one considers as the most important ones, although not as rational ones. Here we only remark that the program leaves room for both ways and, therefore, it cannot be criticized to be restricted to (cognitive and) instrumental rationality.

## 2. COGNITIVE RATIONALITY AND PATTERNS OF EXPECTATION

Expectations are related to a certain situation, a certain context. We may have expectations with respect to the future course of events (in a context) as well as with respect to the true description of the objects and their relations. Using terms in a loose way we may say that our expectations are always directed to theories and (outcomes of) experiments within a context of which the boundaries are given by the knowledge we have or presuppose. In Chapter 4 we shall give a sharp definition of a so-called paradigmatic context, but for our present purposes we do not need that idealization. Here we confine ourselves to a number of remarks.

Theories need not be of a universal nature, i.e., they may be of an existential nature and even be restricted to one individual object, e.g., the theory that there is life on Mars. On the other hand, experiments may be construed in such a way that the outcome of one experiment is of a universal nature, e.g., the outcome that a particular type of gas behaves such and such. It will be clear that, in this approach, the distinction between theories and outcomes of experiments is to be based on the specific nature of the context. Possibilities with respect to which we can do direct observations are called (possible) outcomes of experiments; if we can do only indirect observations we call them theories.

Suppose, for example, that we can investigate the colours occurring on a ball only indirectly with the aid of some mechanism which shows us only the colour of the points at which the ball comes at rest in successive throws. Then we shall call statements to the effect that such and such colours occur on the ball theories and, of course, the colours of the rest points outcomes of experiments. On the other hand, if we could observe the colours on the ball directly, the former statements would not be called theories but (the description of) outcomes of experiments. In that case the example also illustrates that the outcome of an experiment may be only virtually uncertain, i.e., before we have performed an experiment we do not know the outcome, although there may be, objectively speaking, only one possible outcome, e.g., the colours actually occurring on the ball.

Everyday discourse is unclear about whether beliefs (e.g., the belief that a particular theory is true) are also called expectations. Here we shall not do so, if these beliefs are based, in some way, on other expectations. For example, if we assign a high probability to a theory, and subsequently, conclude to believe that the theory is true, we call the former an expectation (a degree of belief) and the latter not. On the other hand we include, contrary to common usage (certain) knowledge in our expectations. Assuming these modifications, we call the total set of expectations with respect to a particular context an expectation pattern.

In a comparative expectation pattern the fundamental relations are 'more probable than' and 'equally probable'. In a real-valued pattern the real numbers assigned to theories and outcomes may be called degrees of expectation or, simply, probabilities. In this study we shall concentrate on real-valued patterns.

Generally speaking, *the aim of the program for the explication of cognitive rationality* is the search for patterns of expectation that we are willing to call rational. For, intuitively, we are inclined to say that rational expectation patterns have at least some systematization.

Part of such systematization may be based on deductive interrelations. For example, in a comparative pattern we shall require that a deductive consequence of a theory is at least as probable as the theory itself. But, in a lot of contexts there will be reason to introduce arguments of another type, e.g., arguments in favour of application of some (weakened) version of the (classical) principle of indifference.

This difference in types of arguments was already indicated in the previous section in terms of formally and materially rational maxims; in the present context we prefer to speak of principles instead of maxims. For the main part our investigations will take place on the basis of a purely mathematical framework. Therefore, the introduction there of material principles cannot be defended in a strict sense. They have to be considered as principles that may be defensible as materially rational in certain types of contexts for which the mathematical framework is appropriate.

At first sight, it may seem to be inappropriate to associate rational expectation patterns with cognitive rationality, instead of the beliefs based, in a rational way, on such patterns. This will be the subject of the next section.

## 3. INDUCTIVE REASONING AND INDUCTIVE PROBABILITY THEORY

Several philosophers have argued that inductive reasoning has to be analyzed in terms of probabilities and inductive rules of inference or rules of acceptance. In this view inductive (valid) arguments are arguments that are not deductively valid but for which holds that the premises (or evidence) make the conclusion (or hypothesis) acceptable. The construction of rules of acceptance is however a difficult problem, for it appears to be not easy to formulate proposals that satisfy some highly reasonable conditions of adequacy: the set of all acceptable hypotheses has to be consistent and deductively closed. Moreover, as a pragmatical condition, it is required that rules of acceptance are only applied on the basis of the total available evidence.

It is, for example, well-known that the most straightforward rule of acceptance, the rule of high probability, does not satisfy the consistency condition, even in the case that the rational probabilities are objective ones. In general, the rule of high probability prescribes to accept a hypothesis $H$ (and to reject its negation) on the basis of evidence $E$, as soon as $p(H/E)$ is larger than $1 - \epsilon$, where $\epsilon$ is a positive small real number (at least smaller than $\frac{1}{2}$). The so-called lottery-paradox may then be construed by requiring that the number of tickets is large enough to reject the hypothesis that a particular ticket will be the winning ticket and therefore reject all similar hypotheses, whereas, at the same time, the hypothesis that there is a winning ticket needs to be accepted because it has (objective) probability 1. Notice that this paradox arises already *a priori*, i.e., without evidence in a proper sense.

Hintikka and Hilpinen (see, Hilpinen [15]) have modified this rule of acceptance, but this modification seems only to work in a highly idealized context and seems therefore rather *ad hoc*. In Chapter 8, Section 6 we shall pay some attention to their proposal. Here we remark only that we do not want to exclude that these sorts of approach may eventually be successful. Fortunately, even if it appears to be unsuccessful, there is a highly plausible view of Carnap as a sort of way-out.

According to Carnap (see, e.g., [4]), inductive reasoning, and therefore inductive logic, is restricted to the assignment of rational probabilities, for it is not the aim of (pure) science to accept hypotheses but only to confirm or disconfirm, i.e., to increase or decrease, their (posterior) probability on the basis of experiments. It is important to note that this does not exclude that some hypotheses are tested while others are assumed to be true, but now only for the sake of the experiment, for it is always possible to incalculate afterwards the probable character of the background hypotheses. Of course, this view of Carnap is a theoretical one and not a plea to reform practice. However, one might even argue that Carnap's view is, in a very informal sense, accepted: a good scientist needs to leave room for the possibility that he will give up even the most strong empirical beliefs and convictions.

In Carnap's view the probabilities play only a role in applied science and practical affairs, i.e., in contexts in which decisions about actions need to be made. In his treatment of decisions about actions Carnap agrees completely with statistical theory and practice. In each context we have to use the decision rule appropriate for that context. For example, the rule of maximizing expected utility may be appropriate in a context in which there is a set of alternative (i.e., mutually exclusive and together exhaustive) hypotheses $H_i$, a set of alternative actions $A_j$, of which one has to be chosen, and a utility function $u$, assigning real values to the result $R_{ij}$ of performing $A_j$ when $H_i$ is true. The rule prescribes then to choose that action $A_j$ for which the expected utility, $\Sigma_i p(H_i/E)u(R_{ij})$, is maximal. In our terminology the research in this direction belongs to the domain of practical rationality.

Consider the following simple way, based on a suggestion of Hilpinen ([15], Section 8.1), of using the rule of maximizing expected utility also for acceptance decisions about alternative hypotheses. Let $A_i$ be the 'action' of accepting $H_i$ as true. Then it seems reasonable to define $u(R_{ij})$ as 1 if $i = j$ and 0 otherwise. By consequence, the expected utility of $A_i$ is equal to $p(H_i/E)$ and the rule prescribes us then to accept the most probable hypothesis. It is clear that this rule cannot directly be seen as a rule of acceptance because it is highly counter-intuitive to apply it if all probabilities are low (which is certainly possible) and if we are, in addition, not forced to accept at least one hypothesis, i.e., if we may postpone judgement.

In order to escape this criticism of the rule, *qua* rule of acceptance, one may include the action $A_p$ of postponing judgement, with $u(R_{ip}) = 1 - \epsilon, 0 < \epsilon < \frac{1}{2}$, for all $i$. The rule of maximizing expected utility leads now to, what might be called, the selection rule of high probability: accept that member of a set of alternative hypotheses for which the probability is larger than $1 - \epsilon$, postpone judgement if there is none.

This rule has as extreme case, for $\epsilon = \frac{1}{2}$, the rule derived by Hempel ([14], Section 6.4), also discussed and criticized by Hilpinen ([15], Section 8.3) on the basis of a 'relative-content measure of utility' and the rule of maximizing expected utility.

Apart from an exceptional case for $\epsilon = \frac{1}{2}$ (two hypotheses with probability $\frac{1}{2}$) one may accept, according to this selection rule, at most one of the hypotheses. Hence, as long as the set of alternative hypotheses is fixed, the lottery-paradox cannot arise. But it may arise if we apply it successively for each ticket on the hypotheses 'it is the winning ticket' and 'it is not the winning ticket' and, therefore, the rule can hardly be seen as a rule of acceptance for inductive inference.

In our opinion the foregoing problems support Carnap's pessimism with regard to inductive rules of inference. One may also use a general argument for this pessimism. On the one side most modern writers agree that inductive rules of inference cannot be truth-preserving or, to put it in classical terms, that a principle of induction cannot exist (and, consequently, that there is no principle of induction to be justified). On the other side one may argue that being truth-preserving is the most general condition of adequacy for rules of inference. Therefore, inductive rules of inference do not and cannot exist.

For this reason Carnap restricts the definition of inductive logic to the study of rational credence functions, i.e., what we have here called rational expectation patterns. We prefer, however, to speak of inductive probability theory, not only because we do not want to exclude the possibility of successful attempts to construct satisfactory inductive rules of inference, but also, and primarily, because all other parts of logic include rules of inference. In this terminology, *inductive probability theory* is the study of probability measures which seem to be candidates for rational expectation patterns in certain types of contexts. *Deductive* or *objective probability theory* is then of course the study of probability measures which are primarily intended as objective descriptions of physical processes.

In Carnap's view two fundamentally different concepts of probability can be outlined: a logical and a statistical one. It will already be clear that we do not take this position. In our view there will be considerable overlap between the measures studied in objective probability theory and those studied in inductive probability theory.

Among mathematicians, applied probability theory refers to the study of empirical interpretations or models of probability measures, i.e., the study of objective applications. Of course, inductive probability measures may also have objective applications and we shall pay attention to some examples of interpretations in this study. But our primary interests are applications in the sense of rational expectation patterns in concrete contexts for which there is no objective probabilistic description or the precise description is unknown These sort of applications will be called inductive applications.

By consequence (pure) inductive probability theory is a vaguely defined branch of probability theory, whereas the distinction between objective and inductive applications of probability theory is a sharp and fundamental one.

Inductive applications pretend to be explications of rational expectation in particular contexts. Inductive probability theory pretends to provide instruments for such applications. An important part of this book is directed to these instruments.

CHAPTER 3

# LOGICO-MATHEMATICAL PRELIMINARIES

This chapter gives a short survey of the logical, set-theoretical and the main probability-theoretical notions and notations that will be used in the sequel. The theorems to be stated will not be proved because these proofs are easily available in textbooks (e.g., Feller [10,11]). In order to keep the text of the next chapters readable the symbolism to be introduced in this chapter will not be used rigorously: symbolic formulations will be alternated with verbal ones. Moreover we shall repeat some main definitions and theorems in the next chapters in the terminology of that particular part of the text.

## 1. LOGICAL VOCABULARY

In order to specify the vocabulary of (first order) predicate logic we introduce individual variables $z$, $z_1$, $z_2$, ... ranging over some universe of discourse. Let $\varphi$, $\varphi_1$, $\varphi_2$, ..., $\varphi(z)$, $\varphi(z_1, z_2, ...)$ indicate statement-formulas. What a statement-formula is will be made clear in a moment. We get new statement-formulas by adding or inserting symbols, which are abbreviations of certain expressions, in the following way:

| | | |
|---|---|---|
| $\sim \varphi$ | not $\varphi$ | (negation) |
| $\varphi_1 \,\&\, \varphi_2$ | $\varphi_1$ and $\varphi_2$ | (conjunction) |
| $\varphi_1 \vee \varphi_2$ | $\varphi_1$ or $\varphi_2$ (i.e., or both) | (disjunction) |
| $\varphi_1 \rightarrow \varphi_2$ | if $\varphi_1$ then $\varphi_2$ | (implication) |
| $\varphi_1 \leftrightarrow \varphi_2 (\varphi_1 \text{iff} \varphi_2)$ | $\varphi_1$ if and only if $\varphi_2$ | (equivalence) |
| $\forall_z(\varphi(z))$ | for all $z$ $\varphi(z)$ | (universal quantification) |
| $\exists_z(\varphi(z))$ | for some $z$ $\varphi(z)$. | (existential quantification) |

The individual variable $z$ is said to be bound in the statement formulas $\forall_z(\varphi(z))$ and $\exists_z(\varphi(z))$ by a quantifier. If $z$ occurs in $\varphi$ but not bound by a quantifier then it is said to occur free, which may be indicated by writing $\varphi(z)$.

This terminology enables us to clarify the notion of statement-formula

16

in terms of statements: A statement-formula is either a statement or an expression which becomes a statement by replacing all free occurrences of individual variables by individual constants (i.e., names of individuals belonging to the universe of discourse). Finally, a statement may roughly be characterized as an expression of which it makes sense to say that it is true or false.

The following notations, defined in terms of the preceding ones, will be helpful to simplify the formulation of complex statements and statement-formulas:

| | |
|---|---|
| $z_1 \varphi z_2$ | $\varphi(z_1, z_2)$ |
| $z_1 \not\varphi z_2$ | $\sim z_1 \varphi z_2$ |
| $\&_i \varphi_i$ | $(\dots ((\varphi_1 \mathbin{\&} \varphi_2) \mathbin{\&} \varphi_3) \mathbin{\&} \dots)$ |
| $\bigvee_i \varphi_i$ | $(\dots ((\varphi_1 \vee \varphi_2) \vee \varphi_3) \dots)$ |
| $\forall_{\varphi_1(z)} \varphi_2(z)$ | $\forall_z (\varphi_1(z) \rightarrow \varphi_2(z))$ |
| $\exists_{\varphi_1(z)} \varphi_2(z)$ | $\exists_z (\varphi_1(z) \mathbin{\&} \varphi_2(z))$. |

We shall not characterize the semantical notion of logical consequence and the syntactical notion of (logical) derivability because our arguments will be at an elementary level.

As a consequence we shall also omit a formal characterization of notions like exclusiveness and exhaustiveness. The following informal characterizations may suffice. Given certain assumptions (a context), then certain statements are mutually exclusive if at most one of them may be true. They are together exhaustive if at least one of them has to be true.

## 2. SET-THEORETICAL VOCABULARY

Let the individual variables $z$, $z'$, $z_1$, $z_2$, ... range over a given set $S$ and let $Z$, $Z'$, $Z_1$, $Z_2$, ... be variables for subsets of $S$. We shall use the following notations and definitions:

| | | |
|---|---|---|
| $z \in Z$ | $z$ is an element of $Z$ | |
| $Z_1 = Z_2$ | $\forall_z (z \in Z_1 \leftrightarrow z \in Z_2)$ | (identity) |
| $Z_1 \supset Z_2 (Z_2 \subset Z_1)$ | $\forall_z (z \in Z_2 \rightarrow z \in Z_1)$ | (inclusion) |
| $Z_2 \subset Z_1$ | $Z_2$ is a subset of $Z_1$ | |
| $Z = \varnothing$ | $Z$ is empty ($\sim \exists_z (z \in Z)$) | |
| $\{z / \varphi(z)\}$ | the set of all $z$ satisfying $\varphi$ | (abstraction) |
| $Z_1 \cap Z_2$ | $\{z / z \in Z_1 \mathbin{\&} z \in Z_2\}$ | (intersection) |
| $Z_1 \cup Z_2$ | $\{z / z \in Z_1 \vee z \in Z_2\}$ | (union) |

| | | |
|---|---|---|
| $Z_1 - Z_2$ | $\{z/z \in Z_1 \ \& \ z \notin Z_2\}$ | (difference) |
| $\overline{Z}$ | $\{z/z \notin Z\}$ | (complement) |
| $Z_1 \cap Z_2 = \varnothing$ | $Z_1$ and $Z_2$ are disjunct (or, non-overlapping) | |
| $P(Z)$ | $\{Z'/Z' \subset Z\}$. | (powerset) |

Alternative notations for abstraction will be clear from the context. It is to be understood that all notations and definitions that have been introduced are also applicable if $Z$, $Z'$, $Z_1$, $Z_2$, ... are considered as individual variables ranging over $P(S)$.

Let $F$ indicate an arbitrary subset of $P(S)$. The following abbreviations will also be used:

| | | | |
|---|---|---|---|
| $\underset{Z \in F}{\cap} Z$ | $(\cap F)$ | $\{z/\forall_{Z \in F} z \in Z\}$ | (intersection) |
| $\underset{Z \in F}{\cup} Z$ | $(\cup F)$ | $\{z/\exists_{Z \in F} z \in Z\}$. | (union) |

$F$ is said to be *countable or denumerable* if $F$ can be written as $\{Z_i/i \in I\}$, where $I$ is a set of natural numbers. In this case we write also:

$$\cap Z_i \ (\text{or}, \ \underset{i}{\cap} Z_i, \ \text{or}, \ \cap Z_i) \text{ instead of } \cap F, \text{ and}$$
$$\underset{i \in I}{\cup} Z_i \ (\text{or}, \ \underset{i}{\cup} Z_i, \ \text{or}, \ \cup Z_i) \text{ instead of } \cup F.$$

$F$ is called a *partition* of $Z$ if its members are mutually non-overlapping and if their union is equal to $Z$. $F$ is a countable partition of $Z$ if $F$ is moreover countable.

$F$ is called a *σ-algebra* in $Z$ if it is a set of subsets of $Z$ closed under complementation and the formation of countable unions and intersections. or, more formally: $F$ is a subset of $P(Z)$ such that

(a)      if $Z' \in F$ then $Z - Z' \in F$

(b)      if $F'$ is a countable subset of $F$ then $\cup F'$ and $\cap F'$ belong to $F$.

In fact it is sufficient to require in (b) only that $F$ is closed under the formation of countable unions, or only that it is closed under the formation of countable intersections, for the remaining requirement can be deduced.

We get the definition of an algebra by replacing in the definition of σ-algebra 'countable' by 'finite'. It is clear that a σ-algebra is always an algebra.

The following theorems will frequently be used:

If $F$ is an algebra in $Z$ then $Z \in F$ and $\varnothing \in F$.
The powerset of a set $Z$ is a σ-algebra in $Z$.

If $F$ is a $\sigma$-algebra in $Z$ and $Z' \in F$ then $F'$ defined by $\{Z'' \cap Z'/Z'' \in F\}$ is a $\sigma$-algebra in $Z'$.

For the introduction of the notion of a *Cartesian product* we suppose that there is a countable number of sets $S_1$, $S_2$, ... such that, for $i = 1$, 2, ..., the individual variable $z_i$ ranges over $S_i$ and that $Z_i$ is a subset of $S_i$. The finite Cartesian product $Z_1 Z_2 \ldots Z_n$ is now defined as the set of all sequences $z_1 z_2 \ldots z_n$ such that $z_i \in Z_i$, $i = 1, 2, \ldots, n$. Similarly, the infinite Cartesian product $Z_1 Z_2 \ldots$ is defined as the set of sequences $z_1 z_2 \ldots$ such that $z_i \in Z_i$, $i = 1, 2, \ldots$. The following notations will also be used:

$$\prod_{i=1}^{n} Z_i \quad (\text{or}, \ \prod Z_i) \quad Z_1 Z_2 \ldots Z_n; \ Z^n \quad ZZ \ldots Z \ (n \text{ times})$$

$$\prod_{i=1}^{\infty} Z_i \quad (\text{or}, \ \prod Z_i) \quad Z_1 Z_2 \ldots; \quad Z^{\infty} \quad ZZ \ldots$$

Our notation for Cartesian products may not be misinterpreted as a notation for intersection!

Let $Y_n$ indicate a subset of $\prod_{i=1}^{n} S_i$ and $Y_\infty$ a subset of $\prod_{i=1}^{\infty} S_i$. A set $Y_\infty$ is said to depend only on finitely many coordinates (or to be essentially finite) if there is for some $n$ a $Y_n$ such that $Y_\infty = Y_n S_{n+1} S_{n+2} \ldots$. A set $Y_\infty$ is called *measurable* if it belongs to the smallest $\sigma$-algebra in $\bigsqcup_{i=1}^{\infty} S_i$ containing all essentially finite subsets.

Let $F_1$ be a $\sigma$-algebra in $S_1$ and $F_2$ in $S_2$. The *product $\sigma$-algebra* $F$ of $F_1$ and $F_2$ is defined as the smallest $\sigma$-algebra in $S_1 S_2$ containing all products $Z_1 Z_2$, $Z_1 \in F_1$ and $Z_2 \in F_2$. The latter products are called *rectangular elements* of $F$.

### 3. SOME ELEMENTS OF PROBABILITY THEORY

We indicate definitions by 'D' and theorems by 'T'.

(D1)   A real-valued function $p(.)$ on a countable set $S$ is an *elementary probability measure* (or, simply, a probability function) on $S$ if
(a) $p(z) \geq 0$   for all $z \in S$
(b) $\sum_{z \in S} p(z) = 1$.

(D2)     A real-valued function $p(.)$ on a $\sigma$-algebra $F$ in a set $S$ is a
         *probability measure on $F$ in $S$* if
         (a) $p(Z) \geq 0$  for all $Z \in F$
         (b) $p(S) = 1$
         (c) for any countable collection of mutually non-overlapping
             elements $Z_1, Z_2, \ldots$ of $F$: $p(\bigcup_i Z_i) = \sum_i p(Z_i)$.

The expression 'probability measure on $F$ in $S$' is abbreviated to 'probability measure on $S$' if $F$ is the powerset of $S$ or if $S$ is an infinite Cartesian product of sets and $F$ is the set of measurable subsets.

(T1)     If $S$ is a countable set, then an elementary probability measure
         on $S$ determines uniquely a probability measure $p(.)$ on $S$ by
         the definition $p(Z) = {}_{df} \sum_{z \in Z} p(z)$ and, conversely, a probability
         measure on $S$ determines uniquely an elementary probability
         measure $p(.)$ on $S$ by the definition $p(z) = {}_{df} p(\{z\})$.

This theorem enables us to talk about a probability measure corresponding to an elementary probability measure on a countable set and vice versa.

The values assigned to elements and sets by a probability measure are called their *probabilities* according to that measure. For the following theorems we presuppose that $p(.)$ is a probability measure on a $\sigma$-algebra $F$ in a set $S$.

(T2)     If $p(Z') = 1$ for some $Z' \in F$ then $p(.)$ restricted to the subset
         $F'$ of $F$ defined by $\{Z \cap Z'/Z \in F\}$, which is a $\sigma$-algebra in $Z'$,
         is a probability measure on $F'$ in $Z'$, coinciding with $p(.)$ as
         far as $F'$ is concerned.

(T3)     Let $Z'$ be an element of $F$ for which $p(Z') \neq 0$ then the real-
         valued function $p(./Z')$ or $p_{Z'}(.)$ defined by $p(Z' \cap Z)/p(Z')$
         is a probability measure on $F$ in $S$; moreover, this function
         restricted to the $\sigma$-algebra $F'$ defined in (T2) is also a pro-
         bability measure on $F'$ in $Z'$.

(T4)     $p(./S) = p(.)$.

The probability measure $p(./Z')$ is called the *conditional probability measure* based on $Z'$ according to $p(.)$ and the value $p(Z/Z')$ is called the *conditional probability* of $Z$ given $Z'$ according to $p(.)$. Though the conditional probability measure is undefined in case $p(Z') = 0$ we then take $p(Z')p(Z/Z')$ to be equal to 0 by convention if this expression occurs in a summation or product.

(T5)   Let $Z, Z', Z_1, Z_2, Z_3, \ldots$ be elements of $F$, then:

(1) $p(Z) = 1 - p(\bar{Z})$;

(2) $p(Z \cup Z') = p(Z) + p(Z') - p(Z \cap Z')$;

(3) $p(\cap_i Z_i) = p(Z_1) \cdot \prod_{\substack{i \neq 1}} p(Z_i / \underset{j=1}{\overset{i-1}{\cap}} Z_j)$

$$= \prod_i p(Z_i / \underset{j=1}{\overset{i-1}{\cap}} Z_j) \quad (\text{where } p(Z_1 / \underset{j=1}{\overset{0}{\cap}} Z_j) = \text{df} \, p(Z_1));$$

(4) If the $Z_i$ are mutually non-overlapping and if $\Sigma_i p(Z_i) = 1$ then

(a) $p(.) \equiv \Sigma_i p(Z_i) p(./Z_i)$   and

(b) $p(Z_i/Z) = p(Z_i) \cdot p(Z/Z_i)/\Sigma_j(p(Z_j) \cdot p(Z/Z_j))$ for all $Z \in F$.

T5(2) is usually called the summation rule, T5(3) the product rule and T5(4b) Bayes' rule or Bayes' theorem.

(T6)   Let $S$ be the Cartesian product $S_1 S_2$, let $Z_1$ indicate a subset of $S_1$ and $Z_2$ a subset of $S_2$, let $F_1$ indicate the $\sigma$-algebra in $S_1$ defined by $\{Z_1/Z_1 S_2 \in F\}$ and $F_2$ the one in $S_2$ defined by $\{Z_2/S_1 Z_2 \in F\}$ then:

(1) $p(Z_1)$ defined by $p(Z_1 S_2)$ is a probability measure on $F_1$ in $S_1$:

(2) for a fixed $Z_1 \in F_1$, for which $p(Z_1 S_2) \neq 0$, $p(Z_2/Z_1)$, or $p_{Z_1}(Z_2)$, defined by $p(S_1 Z_2/Z_1 S_2)$ is a probability measure on $F_2$ in $S_2$:

if, in addition, $F$ is the product $\sigma$-algebra of $F_1$ and $F_2$, then

(3) $p(.)$ is completely determined by the probabilities of the rectangular elements of $F$;

(4) $p(.)$ is completely determined by the probability measure on $F_1$ in $S_1$ defined under (1) and the probability measures on $F_2$ in $S_2$ defined under (2) for all $Z_1 \in F_1$ for which $p(Z_1) \neq 0$.

(D3)   Let $S_1, S_2, S_3, \ldots$ be a countable sequence of sets and let $R_n$ indicate the Cartesian product $\prod_{i=1}^{n} S_i$ then a sequence $p_1, p_2, p_3, \ldots$ of probability measures on $R_1, R_2, R_3, \ldots$ is called a *consistent probability pattern* w.r.t. $R_1, R_2,$

$R_3, \ldots$ if for all $n$ and subsets $Y_n$ of $R_n$ holds $p_n(Y_n = p_{n+1}(Y_n S_{n+1})$.

(T7)  A consistent probability pattern $p_1$, $p_2$, $p_3$, ..., $P_N$ w.r.t. $R_1$, $R_2$, ..., $R_N$, based on a finite sequence of sets $S_1$, $S_2$, ..., $S_N$, is completely determined by $p_N$ (because $p_n(Y_n) = p_N(Y_n S_{n+1} S_{n+2} \ldots S_N)$ holds generally).

(T8)  Let $p_1$, $p_2$, ... be a consistent probability pattern w.r.t. $R_1$, $R_2$, ... based on a denumerably infinite sequence of sets $S_1$, $S_2$, ... and let $R_\infty$ indicate the infinite Cartesian product $\Pi_i S_i$ then:
(1) there is at most one consistent extension of the pattern to $R_\infty$, i.e., there is at most one probability measure $p(.)$ on (the set of measurable subsets of) $R_\infty$ for which generally holds $p(Y_n S_{n+1} S_{n+2} \ldots) = p_n(Y_n)$;
(2) if the sets $S_1$. $S_2$, ... are countable then there exists such a (unique) consistent extension. for which holds. moreover:
(a) it is completely determined by the definition
$p(Y_n S_{n+1} S_{n+2} \ldots) =_{\mathrm{df}} p_n(Y_n)$ and
(b) it determines the whole pattern (because of this definition).

T8 is known as the theorem of Kolmogorov, restricted to countable sets for the existence-claim. Among other things it enables us, together with T7, to indicate a consistent probability pattern w.r.t. $R_1$, $R_2$, ... based on a countable sequence of *countable* sets $S_1$, $S_2$, ... unambiguously by $p$ or $p(.)$.

Let $p$ indicate such a consistent probability pattern based on countable sets. Because of T1 we may interchange $\{z_1 z_2 \ldots (z_n)\}$ and $z_1 z_2 \ldots (z_n)$ without ambiguity if one of these expressions occurs explicitly as argument or as accepted condition in a probability-expression.

(T9)   (1) $p(Y_n) = \displaystyle\sum_{z_1 z_2 \ldots z_n \in Y_n} p(z_1 z_2 \ldots z_n)$.

(2) $p(Z_1 Z_2 \ldots Z_k Z_{k+1} \ldots Z_n)$
$= p(Z_1 Z_2 \ldots Z_k) \cdot p(Z_{k+1} \ldots Z_n / Z_1 Z_2 \ldots Z_k)$
$= \displaystyle\sum_{z_1 z_2 \ldots z_k \in Z_1 Z_2 \ldots Z_k} p(z_1 z_2 \ldots z_k) \cdot$
$\displaystyle\sum_{z_{k+1} \ldots z_n \in Z_{k+1} \ldots Z_n} p(z_{k+1} \ldots z_n / z_1 z_2 \ldots z_k)$

$$= \sum_{z_1 z_2 \ldots z_k z_{k+1} \ldots z_n \in Z_1 Z_2 \ldots Z_k Z_{k+1} \ldots Z_n}$$

$$p(z_1 z_2 \ldots z_k) p(z_{k+1} \ldots z_n / z_1 z_2 \ldots z_k).$$

(3) $p(Z_1 Z_2 Z_3 \ldots Z_n)$

$$= p(Z_1) p(Z_2 / Z_1) p(Z_3 / Z_1 Z_2) \ldots p(Z_n / Z_1 Z_2 Z_3 \ldots Z_{n-1})$$

$$= \sum_{z_1 \in Z_1} p(z_1) \sum_{z_2 \in Z_2} p(z_2 / z_1) \sum_{z_3 \in Z_3} p(z_3 / z_1 z_2) \ldots$$

$$\sum_{z_n \in Z_n} p(z_n / z_1 z_2 \ldots z_{n-1})$$

$$= \sum_{z_1 z_2 z_3 \ldots z_n \in Z_1 Z_2 Z_3 \ldots Z_n}$$

$$p(z_1) p(z_2 / z_1) p(z_3 / z_1 z_2) \ldots p(z_n / z_1 z_2 z_3 \ldots z_{n-1}).$$

# CHAPTER 4

# FORMALLY RATIONAL EXPECTATION IN A PARADIGMATIC CONTEXT

## 1. Paradigmatic contexts

This book is essentially restricted to expectation in a particular type of context. It is concerned with contexts corresponding to clear-cut countable scientific situations, i.e., countable situations in which the conceptualization of the relevant part of the world as well as the deductive interrelations are unproblematic. We call this type of context paradigmatic because it is often considered to represent the ideal or, at least, the simplest type of genuine scientific problem-situations. It does not seem, however, that such situations occur frequently in pure science nor that they constitute the most important universe of discourse for the philosophy of science. On the other hand one might argue that such situations occur frequently in practical affairs, such as applied sciences, and that they are therefore a topic of some interest for a general theory of the applied sciences.

What has been said so far applies also to clear-cut situations that do not have the property of countability but, for simplicity, we do not include such situations in the present study because it seems rather difficult, although not impossible, to extend the analysis of this study to these types of clear-cut situations.

As explication of the notion of *paradigmatic context* we assume that there is:

—a (physical) system and a(n) (interpreted) language;

—a countable set of *elementary* (i.e., mutually exclusive and together exhaustive) *theories* related to the system;

—a countable sequence of experiments with respect to the system;

—for each experiment a countable set of elementary (i.e., mutually exclusive and together exhaustive) descriptions of the outcomes, or, for short, a set of *elementary outcomes*;

—in addition, it is assumed that the formulation of theories and outcomes gives rise to a well-defined *relation of compatibility* between the elementary theories and the sequences of elementary outcomes.

A few remarks about the reach of this explication are needed. First, it has not been assumed that the system is not changed by the experiments: we permit such, and only such, changes but the elementary theories are then supposed to be related to the original system. Second, each experiment may in fact be a (countable) set of physically different experiments, provided that the program for the selection of the experiment actually to be performed at a certain stage is specified beforehand. This program may or may not be probabilistic and it may or may not base the selection on the outcomes of the foregoing experiments. Finally, the compatibility relation need not be strictly logical, e.g., certain infinite sequences of elementary outcomes may be logically not excluded by an elementary theory, but still be incompatible in the sense that, given the theory, it is with objective probability 1 (probabilistically certain) that no such sequence will occur.

Though we shall consider in the next chapters some examples of paradigmatic contexts we shall develop in this chapter a general framework for such contexts in order to stress the general and formal character of some conditions of adequacy for rational expectation in such a context. As a consequence this chapter provides also a point of departure for the study of rational expectation in other examples of a paradigmatic context.

## 2. TWO CONDITIONS FOR RATIONAL EXPECTATION

Let $G_N$ indicate a paradigmatic context with a finite number $N$ of experiments and $G$ one with denumerably infinite many experiments. Given a context $G_{(N)}$ (i.e., a finite or an infinite one) we obtain, of course, a new (finite paradigmatic) context if we take only into consideration the first $n(n = 1, 2, \ldots, (N))$ experiments: $G_n$.

Let $W_n(n = 1, 2, \ldots, (N))$ indicate the countable set of elementary outcomes of the $n$-th experiment and let $V_n$ indicate the Cartesian product $W_1 W_2 \ldots W_n$. Let $D$ indicate the set of elementary theories. The Cartesian product $DV_n$ will be called the $n$-space of $G_{(N)}$ and a subset $A_n$ of this $n$-space an $n$-perspective of $G_{(N)}$. The set of $n$-perspectives, i.e., the powerset of $DV_n$, is called the $n$-scope of $G_{(N)}$.

We define an *expectation function* w.r.t. the $n$-space $DV_n$ of $G_{(N)}$ as a real-valued function $f_n$ on the $n$-scope of $G_{(N)}$. An *expectation pattern* w.r.t. $G_{(N)}$ is defined as a sequence of expectation functions $f_1, f_2, \ldots, (f_N)$ w.r.t. $DV_1, DV_2, \ldots, (DV_N)$ respectively. It is important to note that we introduce the (real-valued) functional character of expectation by definition and not as a condition of adequacy based on our intuitive notion of rational expectation. We do not want to exclude the possibility that comparative expectation patterns will be developed (i.e., patterns in terms of the relations 'more probable than' and 'equally probable') that are rational.

We are now in a position to qualify the formulation of the main problem of this study. If we call $f_n(A_n)$ the *degree of expectation* of the $n$-perspective $A_n$ of $G_{(N)}$ according to the expectation function $f_n$ then the subject of our research, rational expectation in a paradigmatic context, might be reformulated more precisely: rational patterns of degrees of expectation w.r.t. the perspectives of a paradigmatic context.

Our first condition of adequacy will not be surprising in the light of what already has been said.

(CA1)   *principle of coherency*
        In order to be rational an expectation pattern has to be probabilistic: the expectation functions constituting the pattern have to be probability measures on the respective scopes.
        i.e., for all $n(n = 1, 2, \ldots, (N))$
        (a) $f_n(A_n) \geq 0$   for all $A_n \subset DV_n$
        (b) $f_n(DV_n) = 1$
        (c) for any countable collection of mutually non-overlapping $n$-perspectives $A_n^1, A_n^2, \ldots : f_n(\cup_i A_n^i) = \Sigma_i f_n(A_n^i)$.

Many people acquainted with probability theory will not need a separate motivation for this condition. For them it is a form of acquired intuition. For others, however, this will not (yet) be the case. Many authors have justified this condition in terms of rational betting behaviour (e.g., Carnap [4]). As a matter of fact coherency has first been defined as a relation between betting quotients. Shimony [33] has proved that a coherent betting system is a probabilistic betting system (and vice versa). Here we shall only try to disentangle the given probabilistic formulation of the coherency condition in terms of basic intuitions, extrapolations, conventions and all their logico-mathematical consequences.

The first basic intuition, in our opinion, is that, in technical terms, rational degrees of expectation are additive, i.e.,

(1)      $f_n(A_n \cup A_n') = f_n(A_n) + f_n(A_n')$   if $A_n \cap A_n' = \varnothing$.

Subcondition (c) of CA1 includes the extensions of (1) to finitely many non-overlapping perspectives as well as to infinitely many. The first extension is a logico-mathematical consequence of (1), the second is not. Because we are, in general, distrustful with regard to our intuitions concerning infinity we prefer to consider the latter extension as an extrapolation (based on good reasons). This completes subcondition (c). That $f_n(DV_n)$ is finite, which is implied by (b), may well be seen as a convention. Together with (1) it implies that $f_n(\varnothing) = 0$, for $DV_n \cup \varnothing = DV_n$.

As a second basic intuition we introduce

(2)     $f_n(A_n) \geq f_n(A_n')$   if $A_n \supset A_n'$.

It implies that $f_n(DV_n) \geq f_n(A_n) \geq f_n(\varnothing)$ for all $A_n$ and therefore, using $f_n(\varnothing) = 0$, subcondition (a). The only thing which remains to be explained is that $f_n(DV_n) = 1$. We shall do so after giving the second condition of adequacy:

(CA2)  *principle of consistency*
       In order to be rational an expectation pattern has to be consistent: for all $n(n = 1, 2, \ldots, (N - 1))$ and $n$-perspectives $A_n$ holds:

(3)     $f_n(A_n) = f_{n+1}(A_n W_{n+1})$.

Notice that the term consistency is not used here in the logical sense, but in the sense of probability theory (see D3 of Chapter 3)

We are inclined to call (3) a basic intuition, but perhaps it is better to call it a convention. In any case, it implies that $f_n(DV_n)$ does not depend on $n$, i.e., it is a fixed (finite positive) number. That the value 1 has been taken (subcondition (b) of CA1) is, of course, entirely conventional.

A few remarks have to be made. First, it is of course possible to specify a different set of basic intuitions. Second, one might start with incompatible intuitions; to choose a consistent subset is a general device of concept explication. Third, the reader who does not share the mentioned intuitions and hesitates to agree with them, may become more conciliatory when he sees the fruitfulness of them.

## 3. A FRAMEWORK FOR A PARADIGMATIC CONTEXT

Before we start the analysis of the two conditions we shall first specify a complete terminological framework for a paradigmatic context, as far as possible without taking the compatibility relation into consideration.

Some duplications with Section 2 are unavoidable. The notion of measurable subset will be used a number of times. It was defined at the end of Section 2 of Chapter 3. Roughly speaking, the measurable subsets of an infinite (Cartesian) product of sets are those subsets which can be constructed on the basis of subsets of finite products.

### 3.1 *Outcomes*

$W_n(n = 1, 2, \ldots, (N))$ indicates the countable set of statements describing the result of the $n$-th experiment in such a way that they are mutually exclusive and together exhaustive. An arbitrary subset of $W_n$ will be indicated by $X_n$ and will be called an *outcome* of the $n$-th experiment. An arbitrary element of $W_n$, as well as the corresponding one-point-set, will be called *elementary outcome* of the $n$-th experiment.

### 3.2. *Evidence*

3.2.1. *Finite Evidence.* $V_n(k)$ indicates the Cartesian product $W_{k+1}W_{k+2}\cdots W_{k+n}$ for $k = 0, 1, \ldots, (N - 1)$ and $n = 1, 2, \ldots, (N - k)$. An arbitrary subset of $V_n(k)$ will be indicated by $E_n(k)$ and will be called *n-evidence after k* or, more generally, *finite evidence after k*. An arbitrary element $x_{k+1}x_{k+2}\cdots x_{k+n}$ of $V_n(k)$, as well as the corresponding one-point-set will be indicated by $e_n(k)$ and will be called *elementary n-evidence after k* or, more generally, *elementary finite evidence after k*.

3.2.2. *Infinite Evidence.* In the case of an infinite context, $V(k)$ indicates the infinite Cartesian product $W_{k+1}W_{k+2}\cdots$. An arbitrary measurable subset of $V(k)$ will be indicated by $E(k)$ and will be called *infinite evidence after k*. An arbitrary element $x_{k+1}x_{k+2}\cdots$ of $V(k)$, as well as the corresponding one-point-set, will be indicated by $e(k)$ and will be called *elementary infinite evidence after k*. (Note that (the one-point-set of) $e(k)$ is measurable for it is the countable intersection of sets $\{x_{k+1}x_{k+2}\cdots x_{k+n}\}V(k + n)$ with $n = 1, 2, \ldots$.)

3.2.3. *A Simplifying Convention.* Up to now $k$ occurred in all notations concerning evidence. From now on we shall omit $k$ if $k = 0$. Hence we get $V_n$, $E_n$, $e_n$ and $V$, $E$, $e$ respectively. We shall also omit in this case the expression 'after $k$' in all names that have been introduced. However, if $k$ occurs unspecified, it remains possible that it is zero. On the other hand if $n$ occurs unspecified, e.g., in $V_n$, then we do not include the possibility that $n$ is infinite, e.g., $V$.

### 3.3. *Theories*
Let $D$ indicate the countable set of statements describing the system in such a way that they are mutually exclusive and together exhaustive. An arbitrary subset of $D$ will be indicated by $T$ and will be called *theory*. An arbitrary element of $D$, as well as the corresponding one-point-set, will be indicated by $t$ and will be called *elementary theory*.

### 3.4. *Perspectives*

3.4.1. *Finite Perspectives.* The Cartesian product $DV_n$ is called the *n-space*. A subset of $DV_n$ will be indicated by $A_n$ and will be called an *n-perspective* or, more generally, a *finite perspective*. An element $te_n$ of $DV_n$, as well as the corresponding one-point-set, will be called an *elementary n-perspective* or, more generally, an *elementary finite perspective*. Finally, the set of $n$-perspectives will be called the *n-scope*.

3.4.2. *Infinite Perspectives.* In the case of an infinite context, $DV$ is called the *space*. A measurable subset of $DV$ will be indicated by $A$ and will be called an *infinite perspective*. The set of infinite perspectives is called the *scope*. Finally, an element $te$ of $DV$, as well as the corresponding one-point-set, will be called an *elementary infinite perspective*.

### 3.5. *Space, Evidence, Perspective and Scope*
If we talk without further specification about space, evidence, perspective and scope these notions are always related to $DV_N$ in case of a finite context $G_N$ and to $DV$ in case of an infinite context $G$.

## 4. First Analysis of a Rational Expectation Pattern

The consequences of the two conditions of adequacy are far-reaching. To begin with: if $G_N$ is a finite context then a consistent expectation pattern w.r.t. it is completely determined by the expectation function $f_N$ for it follows from CA2 that, for all $n$ and $n$-perspectives $A_n$,

$$(4) \qquad f_n(A_n) = f_N(A_n W_{n+1} W_{n+2} \ldots W_N).$$

Now it is also easy to prove that if this $f_N$ is a probability measure on the scope of $G_N$ then the (consistent) expectation pattern is moreover probabilistic in the sense of CA1. Therefore we have that a consistent probabilistic expectation pattern w.r.t. a finite paradigmatic context is completely determined by the corresponding probability measure on its scope. According to our conditions of adequacy it follows now that the determination of rational expectation patterns, as far as finite contexts are concerned, can be considered to be reduced to the determination of rational probability measures on the scope of perspectives or, more loosely, to the assignment of rational probabilities to the perspectives.

In an infinite paradigmatic context $G$ the situation is more complicated. According to Kolmogorov's theorem T8 (Chapter 3), a consistent probabilistic expectation pattern determines, because it is a consistent probability pattern, a unique probability measure on the scope of $G$, by the definition

(5)     $f(A_n V(n)) = f_n(A_n)$.

On the other hand it is easy to see that a probability measure $f$ on the scope of perspectives of $G$ determines by the same definition, but read in the other way, a unique consistent probabilistic expectation pattern $f_1, f_2 \ldots$ w.r.t. $G$.

According to these observations we may generalize the foregoing two conclusions as follows. The first is that a consistent probabilistic expectation pattern w.r.t. a finite or infinite paradigmatic context is completely determined by a probability measure on its scope of perspectives and vice versa. This implies that we may indicate an expectation pattern by the corresponding probability measure. The second conclusion is that the determination of the class of rational expectation patterns w.r.t. a finite or infinite paradigmatic context as a subclass of the consistent probabilistic expectation patterns w.r.t. that context, is equivalent to the determination of the class of rational probability measures on the scope of perspectives of that context or, more loosely, the assignment of rational probabilities to the perspectives.

Now let $f$ be a probability measure on the scope of perspectives. By the following definition

(6.1)     $f(A_n) = _{df} f(A_n V(n))$,          if the context is infinite

(6.2)     $f(A_n) = _{df} f(A_n V_{N-n}(n))$,   if the context is finite

we get back the expectation functions constituting the expectation pattern and they are of course probability measures on the $n$-scopes. The mutual consistency can now be expressed as follows:

(7.1)     $f(A_k V_n(k)) = f(A_k)$,   $k \neq 0$

(7.2)     $f(A_k V(k)) = f(A_k)$,     $k \neq 0$ (only for infinite contexts).

Our probability measure gives rise to a large number of other probability measures. We shall abbreviate 'probability measure on the set of (measurable) subsets of' by 'p.m. on'. In the case of infinite contexts the definitions and theorems (9)–(20) remain adequate/valid if $V_n(k)$, $E_n(k)$, $V_n$, $E_n$ are replaced by $V(k)$, $E(k)$, $V$, $E$ respectively. For finite contexts the upperbound of $k$ is $N - 1$ and that of $n$ is $N - k$. In the definitions (9)–(11) $V_k$ is supposed to vanish in the right side if $k = 0$.

(8.1)     $f(T) = _{\text{df}} f(TV_N)$,    p.m. on $D$, finite context

(8.2)     $f(T) = _{\text{df}} f(TV)$,    p.m. on $D$, infinite context

(9)     $f(E_n(k)) = _{\text{df}} f(DV_k E_n(k))$          p.m. on $V_n(k)$

(10)     $f(T/E_n(k)) = _{\text{df}} f(TV_k E_n(k))/f(E_n(k))$   p.m. on $D$

(11)     $f_T(E_n(k)) = _{\text{df}} f(TV_k E_n(k))/f(T)$          p.m. on $V_n(k)$

(12)     $f(E_n(k)/E_k) = _{\text{df}} f(E_k E_n(k))/f(E_k)$          p.m. on $V_n(k)$

(13)     $f_T(E_n(k)/E_k) = _{\text{df}} f_T(E_k E_n(k))/f_T(E_k)$   p.m. on $V_n(k)$.

The definitions (10)–(13) only make sense, of course, in the case where the denominators at the right side are non-zero. We shall call the p.m. on $D$ defined by (8) the *prior belief function* and the one defined by (10), for $k = 0$ and finite $n$, the *posterior belief function* based on $E_n$. The p.m. on $V_n$ defined by (9), for $k = 0$ and finite $n$, is called the (absolute) *prediction function* (for $n$ experiments) and the one defined by (11), for $k = 0$ and finite $n$, the *T-prediction* function or, more generally, a conditional or relativized prediction function. Finally, the sequence of (*T*-)prediction functions for $n = 1, 2, \ldots, (N)$ is called the (*T*-)*prediction pattern*.

The following theorems, which are easy to prove, relate the different probability measures.

(14)     $f(TV_n) = f(T)$

(15)     $f(T/V_n(k)) = f(T)$

(16)     $f_D(E_n(k)) = f(E_n(k))$

(17)     $f(E_n(k)/V_k) = f(E_n(k))$     $k \neq 0$

(18)     $f_T(E_n(k)/V_k) = f_T(E_n(k))$     $k \neq 0$

(19)      $f_D(E_n(k)/E_k) = f(E_n(k)/E_k)$   $k \neq 0$

(20)      $f_D(E_n(k)/V_k) = f(E_n(k))$.      $k \neq 0$.

The following theorem about prediction functions will be very helpful. Let $E_n$ be equal to the Cartesian product $X_1 X_2 \ldots X_n$ and let $E_k$, for $k = 1, 2, \ldots, n$, be equal to $X_1 X_2 \ldots X_k$ then it is easy to prove that (see T9, Chapter 3)

(21)      $f_T(E_n) = f_T(X_1) f_T(X_2/E_1) f_T(X_3/E_2) \ldots f_T(X_n/E_{n-1})$

provided that $f_T(E_k)$ is unequal to 0 for all $k$. If $f_T(E_k) = 0$ for some $k$ then we have not only that $f_T(E_n) = 0$ but also that one of the factors at the right side of (21) is zero. Hence (21) may be said to hold unconditionally, though by convention.

Now the important question arises what we minimally need to specify about $f$ in order to be able to elaborate $f$ completely with the aid of the theorems of probability. We have already seen that even for an infinite context we may restrict ourselves to (essentially) finite perspectives because of Kolmogorov's theorem. Now it is also easy to prove (T6, Chapter 3) that it is sufficient to determine the values for rectangular perspectives, i.e., perspectives which have the form $TE_n$. Finally we have, because each $n$-space is countable, that

(22)      $f(TE_n) = \sum_{te_n \in TE_n} f(te_n)$

(23)      $f(TE_n) = \sum_{t \in T} (f(t) \cdot \sum_{e_n \in E_n} f_t(e_n)) = \sum_{te_n \in TE_n} f(t) f_t(e_n)$.

From (23) it follows that we are able to calculate $f(TE_n)$ in general as soon as we have:

—$f(t)$ for all $t \in D$, which according to (8) (and T1, Chapter 3) is an elementary probability measure on $D$: *the elementary prior belief function*;

—for all $t$ for which $f(t) \neq 0$: $f_t(e_n)$ for all $n$ and $e_n$ in $V_n$, which according to (11) (and T1, Chapter 3) is, for fixed $n$, an elementary probability measure on $V_n$: *the t-prediction function* (and the sequence: *t-prediction pattern*).

At this point it will appear to be convenient to include the postulated compatibility relation in our framework in order to be able to state the third formal condition of adequacy.

## 5. A FRAMEWORK FOR A PARADIGMATIC
### CONTEXT (CONTINUED)

In this section we complete the framework developed in Section 3 by including the compatibility relation (between elementary theories and sequences of elementary outcomes), which has been assumed to belong to a paradigmatic context.

### 5.1. *Generalization of the Compatibility Relation*
In an infinite context the assumption concerning compatibility amounts to: for all $t$ there is given a measurable subset $V^t$ of $V$ containing precisely all elementary (infinite) evidence compatible with $t$. By $V_n^t(k)$ we indicate the subset of $V_n(k)$ containing those $e_n(k)$ for which there are $e_k \in V_k$ and $e(k + n) \in V(k + n)$ such that $e_k e_n(k)e(k + n)$ belongs to $V^t$.

In a finite context the assumption amounts of course to: for all $t$ there is given a subset $V_N^t$ of $V_N$ containing all elementary $N$-evidence compatible with $t$. By $V_n^t(k)$ we indicate now the subset of $V_n(k)$ containing those $e_n(k)$ for which there are $e_k \in V_k$ and $e_{N-k-n}(k + n) \in V_{N-k-n}(k + n)$ such that $e_k e_n(k)e_{N-k-n}(k + n) \in V_N^t$.

For convenience we shall assume that $V^t$ resp. $V_N^t$ is non-empty for all $t$. The following generalization of the compatibility relation is straightforward. *T is compatible with* $E_n(k)$, abbreviated by $T \circ E_n(k)$, if and only if, the intersection $E_n(k) \cap V_n^t(k)$ is non-empty for some $t$ belonging to $T$. If $T$ is not compatible (in-compatible) with $E_n(k)$ we write $T \not\circ E_n(k)$.

### 5.2. *Subspaces*
Let $W_n^t$ indicate $V_1^t(n - 1), n = 1, 2, \ldots, (N)$. From the assumption that the theories and outcomes are both exhaustive it follows that $W_n = \bigcup_{t \in D} W_n^t$.

We define the *t-space* (for $n$ experiments) $U_n^t$ as the Cartesian product $\prod_{k=1}^{n} W_k^t$. Of course, $V_n^t$ is a subset of $U_n^t$ and it may in fact be a proper subset. The members of $U_n^t$ might be called *weakly compatible* with $t$. Note that $\bigcup_{t \in D} U_n^t$ is a subset of $V_n$ and that it may be a proper subset.

### 5.3 *Kinds of Perspectives*
If $T \not\circ E_n$ we call the perspective $TE_n$ impossible, and if $T \circ E_n$ we call it possible. It is rather natural to extend this to: a perspective $A_n$ is *possible* if there are $t$ and $e_n$ such that $te_n \in A_n$ and $t \circ e_n$, otherwise it is *impossible*.

According to our explication of a paradigmatic context the set of elementary theories as well the sets of elementary outcomes are exhaustive, i.e., at least one (and at most one) of the elementary theories is true and at least one (and at most one) of the elementary outcomes of a particular experiment will be the result of that experiment. This implies that the compatibility relation has to be such that $DV_n$ is a possible perspective and that, because of this implication, we may subdivide the perspectives as follows:

—*impossible perspectives*, as defined above;

—*certain perspectives: $A_n$ is certain if $DV_n - A_n$ is impossible*,

—*contingent perspectives: $A_n$ is contingent if it is possible*, as defined above, but not certain.

### 5.4. *Explication of Some Generally used Expressions*

We define $V_n^D(k) = {}_{df} \bigcup_{t \in D} V_n^t(k)$. Note that $V_n^D$ may be a proper subset of $V_n$. Evidence $E_n(k)$ is said to be *theoretically possible evidence* if $E_n(k) \cap V_n^D(k) \neq \varnothing$. Theory $T$ is *verifiable by a finite number of experiments* if, for some $n$, $V_n - {}_{t \in D-T} \bigcup V_n^t$ is theoretically possible evidence and $T$ is *falsifiable by a finite number of experiments* if $D - T$ is verifiable by a finite number of experiments.

Finally, $T$ is *decidable by a finite number of experiments* if $\bigcup_{t \in T} V_n^t \cap \bigcup_{t \in D-T} V_n^t = \varnothing$ for some $n$. It is easy to check that a decidable theory is verifiable and falsifiable and that the converse does not hold in general.

### 5.5. *Extension to Infinite Evidence and Perspectives*

In case of an infinite context we indicate by $V^t(k)$ of course the subset of $V(k)$ containing all those $e(k)$ for which there is $e_k \in V_k$ such that $e_k e(k) \in V^t$. The following extension of the notions introduced above is rather obvious.

$T$ is compatible with $E(k)$, $T \circ E(k)$, if $E(k) \cap V^t(k) \neq \varnothing$ for some $t \in T$; $U^t = {}_{df} \coprod_{k=1}^{\infty} W_k^t$, $U^t \supset V^t$, $\bigcup_{t \in D} U^t \subset V$; $A$ is possible if there is $te \in A$ such that $t \circ e$, otherwise impossible; $A$ is certain if $DV - A$ is impossible; $A$ is contingent if it is possible but not certain.

$V^D(k) = {}_{df} \bigcup_{t \in D} V^t(k)$, $V^D$ may be a proper subset of $V$; $E(k)$ is theoretically possible if $E(k) \cap V^D(k) \neq \varnothing$; $T$ is verifiable if $V - \bigcup_{t \in D-T} V^t$ is theoretically possible; $T$ is falsifiable if $D - T$ is verifiable; $T$ is decidable if

$\bigcup_{t \in T} V^t \cap \bigcup_{t \in D-T} V^t \neq \varnothing$, again decidability implies verifiability and falsifiability, but not conversely.

Note that a theory may be verifiable/falsifiable/decidable, though not by a finite number of experiments, which might be emphasized by speaking of *countably* verifiable/falsifiable/decidable.

## 6. THIRD FORMAL CONDITION FOR RATIONAL EXPECTATION

The empty set is for each $n$ an impossible $n$-perspective on trivial grounds. In Section 2 we have already seen that the first two conditions of adequacy lead to $f(\varnothing) = 0$. Suppose now that $A_n$ is an impossible non-empty $n$-perspective. This means that this perspective cannot (happen to) be true according to what we know about the context, although our language, in which we describe the system and the experiments, is such that it permits the formulation of this perspective. In our opinion it is reasonable to require that in this case $f(A_n) = f(\varnothing) = 0$. Note that this is equivalent to the requirement that all certain perspectives get the same value as $DV_n$, i.e., 1. Therefore we introduce:

(CA3) *principle of certainty* (*impossibility*)

      In order to be rational an expectation pattern has to assign equal values to all certain (impossible) perspectives: if $A_n$ is certain (impossible), then $f(A_n) = f(DV_n) = 1$ ($f(A_n) = f(\varnothing) = 0$).

A direct consequence of CA3 is

(24)     if $t \varnothing e_n$, then $f(te_n) = 0$ and, if $f(t) = 0$, then $f_t(e_n) = 0$.

Therefore (23) reduces to

(25)     $f(TE_n) = \sum_{t \in T: t \varnothing E_n} {}'f(t) \sum_{e_n \in V_n^t \cap E_n} f_t(e_n)'$.

According to (11) $f_t$ is a consistent probability pattern w.r.t. $V_1, V_2, \ldots,$ ($V_N$). Therefore $f_t(V_n) = 1$. Moreover we have that $f(tV_n) = f(t)$. Now (25) implies that $f(tV_n) = f(tV_n^t)$. It now directly follows that

(26)     $f_t(V_n^t) = 1$

and because $V_n^t$ is a subset of $U_n^t$ we have also

(27)     $f_t(U_n^t) = 1$.

Hence we may conclude that the $t$-prediction pattern $f_t$ can also be considered as a consistent probability pattern w.r.t. $U_1^t, U_2^t, \ldots, (U_N^t)$. Necessary and sufficient for (26) is that $f_t(V^t) = 1$ in case of an infinite context and that $f_t(V_N^t) = 1$ in case of a finite context. A consistent probability pattern w.r.t. $U_1^t, U_2^t, \ldots, (U_N^t)$ satisfying this condition is said to be *closed for t*.

The conclusion at the end of Section 4 may now be reformulated as follows. A rational expectation pattern $f$ is completely determined by the (elementary) *prior belief function* and, for all $t$, the *t-prediction patterns* which are closed consistent probability patterns w.r.t. $U_1^t, U_2^t, \ldots, (U_N^t)$.

The three conditions of adequacy that have been introduced in this chapter could be stated without reference to material (or factual) assumptions about the system and the experiments: these three conditions are therefore called *formal* conditions of adequacy. In our opinion there are no other conditions which can be argued for without assuming particular postulates about the context. Therefore, we call an expectation pattern w.r.t. a paradigmatic context *formally rational* as soon as it satisfies the conditions CA1, CA2 and CA3.

## 7. Decidable contexts

A (countably) *decidable* context is a paradigmatic context in which all elementary theories are decidable in the sense defined in Sections 5.4 and 5.5. We shall first consider a decidable infinite context. In such a context we have, for all $t \neq t'$, $V^t \cap V^{t'} = \varnothing$. We generalize the notion of $t$-space by $U_n^t(k) = \prod_{r=k+1}^{n} W_r^t$. Let $V^t e_n$ be short for $V^t \cap e_n V(n)$. Note that $V^t e_n = V^t \cap e_n U^t(n) = V^t \cap e_n V^t(n)$. Let $f$ be a formally rational expectation pattern. Then it follows that $f$ is completely determined by the (absolute) prediction pattern defined by (9), for it is possible to prove that

$$(28) \qquad f(te_n) = f(V^t e_n).$$

*Proof:* $f(te_n) = f(te_n V(n)) = f(te_n U^t(n)) = f(t(V^t \cap e_n U^t(n)))$
$= f(D(V^t \cap e_n U^t(n))) = f(V^t \cap e_n U^t(n)) = f(V^t e_n).$

Direct consequences of (28) are

$$(29) \qquad f(t/e_n) = f(V^t/e_n) \quad (= {}_{df} f(V^t e_n)/f(e_n))$$

$$(30) \qquad f_t(e_n) = f(e_n/V^t) \quad (= {}_{df} f(V^t e_n)/f(t)).$$

Suppose that there are sets $H_t$ for all $t \in D$ constituting a partition of $V$ such that $U^t \supset H_t \supset V^t$. It is easy to check that all these results remain valid if we replace $V^t$ systematically by $H_t$.

Finally, the situation in a decidable finite context is the same: we need only to replace $V^t$ by $V_N^t$, $V$ by $V_N$, $U^t(n)$ by $U_{N-n}^t(n)$ and $H_t$ by $H_t(N)$.

Therefore we may generally conclude that in the case of a decidable context a rational expectation pattern is completely determined by the (absolute) prediction pattern, which is a consistent probability pattern w.r.t. $V_1, V_2, \ldots, (V_N)$ such that $\sum_{t \in D} f(V_{(N)}^t) = 1$.

By consequence, in case of a decidable context we may approach the specification of a formally rational expectation pattern from two sides: either by the specification of the prior belief function and the conditional prediction patterns or by the specification of the absolute prediction pattern. Both types of prediction patterns are consistent probability patterns w.r.t. the (sequences of) outcomes of the experiments. The conditional ones have to be closed for the corresponding elementary theories.

In this book we shall restrict our attention to decidable contexts in which the set of elementary outcomes is the same for all experiments and in which the same is true for all outcomes compatible with a particular theory. The following two chapters are devoted to consistent probability patterns based on a fixed set of elementary outcomes. These patterns may be used as absolute prediction patterns and, as far as they are closed w.r.t. particular theories, as conditional prediction patterns.

CHAPTER 5

# GENERALIZED CARNAPIAN SYSTEMS

## 1. Introduction

Carnap has always considered the problem of explication of rational (degree of) belief as the problem of the determination of the 'logical probability' of a hypothesis in the light of evidence (see, e.g., [2] and [7]). However, he does not make assumptions about the way in which evidence is obtained. More precisely, Carnap presupposes that the logical probability can be determined as soon as the non-logical constants of the language in which hypothesis and evidence are formulated are known. A consequence of this approach is that Carnap's famous continuum of inductive methods for monadic predicate language (Carnap [3], see also Kemeny [21]) is a purely mathematical probability system presented in the form of a particular application.

Although our approach to rationality will include considerations about the way in which evidence has been obtained, the formal system behind Carnap's continuum will appear to be of great value to us. In Chapter 7 we shall argue that this formal system fits very well in certain kinds of paradigmatic contexts.

What has been said about Carnap's system applies also to the generalization of it presented by Stegmüller [34]. In this chapter we shall derive and analyse this latter system which we shall call a generalized Carnapian system or GC-system. We shall also present in this chapter some of its mathematical interpretations. These interpretations will clarify what it means to apply a GC-system.

## 2. Constitutive principles and definition of GC-systems

Let W be a countable set containing the elements $Q_1, Q_2, \ldots, (Q_w)$. (An expression between brackets at the end of a sequence is supposed to vanish if the sequence is denumerably infinite.) We indicate a particular element $Q_{i_1} Q_{i_2} \ldots Q_{i_n}$ of $W^n$, the $n$-th Cartesian product of $W$, by $e_n$ and a particular subset of $W^n$ by $E_n$.

In Chapter 3 Section 3 we have introduced the notion of a consistent

38

probability pattern. If $p$ is such a pattern with respect to the sequence of sets $W, W^2, W^3, \ldots, (W^N)$ then it is a real-valued function on the elements and subsets such that, for $n$ is $1, 2, \ldots, (N)$,

(1.1) $\quad p(e_n) \geq 0$ $\qquad$ (1.2) $\quad \Sigma\, p(e_n) = 1$

(2) $\qquad p(E_n) = \sum_{e_n \in E_n} p(e_n)$

(3) $\qquad p(E_n) = p(E_n W)$ for $n < N$.

According to Kolmogorov's theorem we have, moreover, that the definition

(4) $\qquad p(E_n WWW \ldots) =_{\mathrm{df}} p(E_n)$

determines a unique probability measure on the set of measurable subsets of the infinite Cartesian product $WWW \ldots$ if $N$ is infinite.

For $p(e_n) \neq 0$ we define the so-called *special values* as follows

(5) $\qquad p(Q_i/e_n) =_{\mathrm{df}} p(e_n Q_i)/p(e_n)$.

If $p(e_n) = 0$ it follows from (1.1), (2) and (3) that $p(e_n Q_i) = 0$ for all $i$. Hence, irrespective of the way in which we define $p(Q_i/e_n)$ in the case $p(e_n) = 0$ as a finite number, we have

(6) $\qquad p(e_n Q_i) = p(e_n)p(Q_i/e_n)$.

The product rule, i.e., repeated application of (6), shows that $p$ is completely determined by the special values, including the initial ones $p(Q_i)$. We replace $p(Q_i)$ by $\gamma_i$ and from (1) we obtain directly

(7.1) $\quad \gamma_i \geq 0$ $\qquad$ (7.2) $\quad \Sigma \gamma_i = 1$.

By induction, from (1) and (6), it can easily be proved that

(8.1) $\quad p(Q_i/e_n) \geq 0$ $\qquad$ (8.2) $\quad \Sigma p(Q_i/e_n) = 1$

if $p(e_n) > 0$ (and $n < N$). In what follows we shall suppose that, in case $p(e_n) = 0$, the special values are defined such that (8) holds also in this case. As far as we can see there is in the present context only one interesting system based on special values in which (8.1) is only satisfied if $p(e_n)$, calculated according to (6), is positive. That system, called the hypergeometric system, will be studied in Section 5.3.

So far we assumed only that $p$ is a consistent probability pattern and we have seen that, if we want to construct such a pattern on the basis of special values, then (7) and (8) represent sufficient and (almost) necessary restrictions to the special values. Now we shall introduce a number of assumptions about $p$ that appear to be the constitutive principles of the generalized Carnapian systems.

Let $n_i(e_n)$ or, for short, if misunderstandings are excluded, $n_i$ indicate the number of occurrences of $Q_i$ in $e_n$ and let $\ddot{e}_n$ indicate the subset of $W^n$ containing all $e'_n$ for which $n_i(e'_n) = n_i(e_n)$ for all $i$. Note that $\ddot{e}_n$ contains $e_n$.

Some of the principles that will be introduced are so-called indifference-principles (I-principles), which may be considered as applications of the methodological principle of indifference, MPI. The latter principle prescribes to assign equal probabilities to a number of possibilities if there seems no reason to assign different ones. Here we talk about an I-principle if it is formally such that it can be an application of MPI in a concrete context.

Suppose for a moment that we impose the following two principles

(POI)    *principle of order indifference*
$$p(Q_iQ_j/e_n) = p(Q_jQ_i/e_n)$$

where $P(Q_iQ_j/e_n)$ is of course defined as $p(Q_i/e_n) \cdot p(Q_j/e_nQ_i)$, which is equal to $p(e_nQ_iQ_j)/p(e_n)$ if $p(e_n) \neq 0$. POI is supposed to include $p(Q_iQ_j) = p(Q_jQ_i)$.

(POI')    $p(Q_i/e_n) = p(Q_i/e'_n)$   if $e'_n \in \ddot{e}_n$.

Consider on the other hand

(SPOI)    *strong principle of order indifference*
$$p(e_n) = p(e'_n)   \text{ if } e'_n \in \ddot{e}_n.$$

These three principles are related according to

(T1)       In a consistent probability pattern w.r.t. $W, W^2, W^3, \ldots, (W^N)$ POI and POI' imply together SPOI (and vice versa for $p(e_n) \neq 0$).

*Proof:*  That SPOI implies POI and POI' for $p(e_n) \neq 0$ is easy to check. The proof of the converse, however, is more complicated. Note first that POI implies

(*)       $p(e_nQ_iQ_j) = p(e_nQ_jQ_i)$

if $p(e_n) \neq 0$ and that (*) holds trivially if $p(e_n) = 0$. Now SPOI is trivial for $n = 1$ and it follows trivially from POI for $n = 2$. As inductive hypothesis we assume that SPOI holds for some $n \geq 2$. Let $e'_{n+1} \in \ddot{e}_{n+1}$. If the last two members of $e_{n+1}$ and $e'_{n+1}$ have at least one member in common, i.e., if $e_{n+1} = e_{n-1}Q_iQ_j$ and $e'_{n+1} = e'_{n-1}Q_iQ_k$, to take the most complicated case, we proceed as follows. By (*) we have $p(e_{n+1}) = p(e_{n-1}Q_jQ_i) = p(e_{n-1}Q_j) \cdot p(Q_i/e_{n-1}Q_j)$. The last product is, on the basis of the inductive hypothesis and POI′, equal to $p(e'_{n-1}Q_k) \cdot p(Q_i/e'_{n-1}Q_k) = p(e'_{n-1}Q_kQ_i)$ which is, because of (*), equal to $p(e'_{n+1})$. Suppose now, on the other hand, that the last two members of $e_{n+1}$ and $e'_{n+1}$ have no common member, i.e., let $e_{n+1} = e_{n-1}Q_iQ_j$ and $e'_{n+1} = e'_{n-1}Q_kQ_l$. Then there needs to be a permutation of $e'_{n+1}$ of the form $e''_{n-1}Q_iQ_l$ for which, on the basis of the inductive hypothesis and POI′, $p(e''_{n-1}Q_iQ_l) = p(e'_{n+1})$. This permutation of $e'_{n+1}$, compared with $e_{n+1}$, brings us back to the first case, $Q.E.D.$

Let us now assume that SPOI holds and let $\gamma_i = 0$ for some $i$. Then it follows by the product rule that $p(e_n) = 0$ as soon as $e_n$ starts with $Q_i$. From SPOI it follows that $p(e_n) = 0$ as soon as $n_i > 0$. (Note that this implies $p(Q_i/e_n) = 0$ if $p(e_n) > 0$, and the latter condition was proved to presuppose $n_i(e_n) = 0$.) From this observation we may conclude that we know what happens with the values for $e_n$ in cases in which the following mathematically convenient, assumption is not fulfilled:

(PIP)    *principle of initial possibility*
         $\gamma_i > 0$.

Consider now the following far-reaching restriction on the possible dependencies of the special values.

(PRR)    *principle of restricted relevance*
         $p(Q_i/e_n) = f_i(n, n_i)$

or, reformulated as an I-principle:

if $e_n$ and $e'_n$ are such that $n_i = n'_i$, then $p(Q_i/e_n) = p(Q_i/e'_n)$.

Note that PRR does not exclude that the special values depend on the 'frame-variables' $w$ and $N$.

It is easy to see that PRR implies POI′. Hence, by T1, if we impose PRR and POI, then SPOI is also satisfied and, therefore, the motivation for PIP remains.

The three principles POI, PRR and PIP appear to be sufficient to derive the GC-systems, provided that $w \neq 2$. In case $w = 2$, however, PRR is not a genuine restriction to $n_i$, for it makes $p(Q_i/e_n)$ implicitly dependent on the number of occurrences of the other element in $W$, viz. $n - n_i$. To derive the GC-systems for $w = 2$ we have to assume in addition that $p(Q_i/e_n)$ is a linear function of $n_i$:

(SPL)    *special principle of linearity* (for $w = 2$)
         if $w = 2$ then $p(Q_i/e_n) = f_i(n, 0) + n_i \cdot g_i(n)$.

If $W$ is a finite set, then it follows from PIP that there is an $i$ for which $0 < \gamma_i \leq \gamma_j$ for all $j$. We indicate this smallest initial special value by $\gamma_{\min}$. If $W$ is (denumerably) infinite then the largest lower-bound of the $\gamma_i$'s is 0 according to (7), in which case we put $\gamma_{\min} = 0$.

In the Appendix to this chapter we shall prove the following basic theorem

(T2)    If a consistent probability pattern with respect to $W$, $W^2$, $W^3, \ldots, (W^N)(N \geqslant 2)$ satisfies the principles POI, PRR, PIP and, if $w = 2$, in addition SPL, then there is a real number $\alpha$ such that

(9)    $0 \leq \alpha \leq (N - 1)/(N - 1 - \gamma_{\min})$   (parameter-condition)
        and such that, replacing $\alpha$ by $\lambda/(1 + \lambda)$, for all $n < N$,

(10)   $p(Q_i/e_n) = f_i(n, n_i) = (n_i + \gamma_i\lambda)/(n + \lambda)$.

The initial special value $p(Q_i) = \gamma_i$ is included in (10) (viz. if $n_i = n = 0$) except if $\lambda = 0$, in which case we impose the inclusion by definition. The parameter-condition is easily seen to be equivalent to the following particular values and intervals for $\lambda$:

(+)    $0 < \lambda < \infty$   $(0 < \alpha < 1)$   or,
        replacing $\lambda$ by $-\eta$ and $(N - 1)/\gamma_{\min}$ by $\eta_{\min}$,

(−)    $\eta_{\min} \leq \eta < \infty$   $(\eta_{\min}/(\eta_{\min} - 1) \geq \alpha > 1)$   or,

($\alpha = 0$) $\lambda = 0$   $(\alpha = 0)$   or,

($\alpha = 1$) $\lambda = \pm\infty$   $(\alpha = 1)$.

It is also easy to see that the interval specified by (−) is non-empty only if $w$ and $N$ are finite. If we talk about negative values of $\lambda$ we suppose that these two conditions are satisfied.

The formulation of (10) corresponds best to the interval (+) though it contains certainly the other cases. If $\alpha = 0$ then (10) reduces to $p(Q_i/e_n) = n_i/n$ and if $\alpha = 1$ to $p(Q_i/e_n) = \gamma_i$. For the interval $(-)$ (10) becomes

$$(10-) \quad p(Q_i/e_n) = (\gamma_i \eta - n_i)/(\eta - n).$$

On the basis of our theorem we introduce the following:

DEFINITION: a Generalized Carnapian system (GC-system) is a consistent probability pattern w.r.t. $W$, $W^2$, $W^3$, ..., $(W^N)$ of a countable set $W$, for which there are real numbers $\gamma_i > 0$ and $\lambda$, such that (10) holds for all $i$ and $n < N$.

From the proof of T2 in the Appendix it can be seen that the parameter-condition (9) is contained in this definition. It is easy to see from (10) that the principles POI and PRR are also contained in this definition, whereas the principle PIP is explicitly mentioned. Moreover, we may conclude from (10) not only that SPL is included but also

(GPL) *general principle of linearity*
$$p(Q_i/e_n) = f_i(n, 0) + n_i \cdot g_i(n).$$

In other words GPL is, according to the theorem, implied by the other principles in case $w \neq 2$, even the stronger version with $g_i(n) = g(n)$.

In the following sections we shall analyse particular classes of GC-systems. In Section 3 we shall consider the whole class, the extreme cases $\lambda = \alpha = 0$ (extreme-inductive GC-systems) and $\lambda = \pm \infty$ ($\alpha = 1$; non-inductive GC-systems), and finally the class of proper Carnapian systems (C-systems) in which $W$ is finite and all $\gamma_i$ are equal to $1/w$. In Section 4 we shall study the class of GC-systems for which $0 < \lambda < \infty$ (positive inductive GC-systems). Finally, Section 5 is concerned with GC-systems for which $\lambda < 0$ (negative inductive GC-systems).

In the analysis we shall pay particular attention to *interpretations* or *models* for GC-systems. At other places in this study we are interested not so much in interpretations but in *applications* of GC-systems. As general characterizations of the notions of interpretation, or model, and application, the following remarks will suffice for the present.

Let there be an experimental process, i.e., a collection of experiments such that for each experiment there is given the set of elementary outcomes. If we (presuppose to) know so much about the process that we can derive the adequate or objective probabilistic description of that process we say that that process provides an interpretation or model for that probabilistic description. On the other hand if we do not know (the details of) the objective probabilistic description of this process we may be interested in the question whether it is possible to construct one in such a way that the resulting probabilistic description can be seen as explication of the intuitive notion of a rational pattern of expectations with respect to that process. In this case we say that the probabilistic description is applied to, or used as, an expectation pattern with respect to that process. Of course, if we know the adequate probabilistic description of a process we choose that description as the (basis for our) expectation pattern.

In the present context of GC-systems the relevant experimental processes have to consist of a number of ($N$) successive experiments such that the set of elementary outcomes ($W$) is the same for all experiments.

In order to prove that such an experimental process is a model for a particular GC-system we need to show that the objective probabilistic description $P$ is a consistent probability pattern w.r.t. $W$, $W^2$, ..., ($W^N$). Moreover we need to show that it coincides with that GC-system. A sufficient way to prove this is, for example, to show that there are positive $\gamma_i$ and a $\lambda$ such that $P$ satisfies (10).

On the other hand if we want to show that a GC-system may be applied to such an experimental process we need first to make plausible that a rational expectation pattern w.r.t. that process has to be a consistent probability pattern. This has been done in general in Chapter 4. One way to complete the justification is to make plausible that the principles POI, PRR and PIP are acceptable. (In the case $|W| = 2$ we have to make plausible in addition that SPL is acceptable.) Another way to complete the justification is to argue that for the purpose of an expectation pattern the process is sufficiently comparable with a process which is known to be a model for GC-systems.

If we have reached the conclusion that we want to use a GC-system as expectation pattern then we still have to determine the values $\gamma_i$ (positive) and $\lambda$ (or $\alpha$ or $\eta$) such that the $\gamma_i$'s sum up to 1 and such that the parameter-condition is fulfilled.

## 3. GENERAL ANALYSIS OF GC-SYSTEMS

### 3.1. *Some Direct Consequences*

Throughout this section we consider an arbitrary GC-system. From (10)

**and** (6) it follows that

(11) $$p(e_n) = \frac{\prod\limits_{i:n_i \neq 0} \gamma_i\lambda \cdot (1 + \gamma_i\lambda)\ldots(n_i - 1 + \gamma_i\lambda)}{\lambda \cdot (1 + \lambda)\ldots(n - 1 + \lambda)}$$

The number of elements of $\ddot{e}_n$ is of course equal to $n!/\Pi n_i!$. Now it follows directly from (2) and the principle of order indifference:

(12) $$p(\ddot{e}_n) = (n!/\Pi n_i!) \cdot p(e_n)$$

and therefore we have in combination with (11)

(13) $$p(\ddot{e}_n) = \frac{\prod\limits_{i:n_i \neq 0} (\gamma_i\lambda/1) \cdot ((1 + \gamma_i\lambda)/2)\ldots((n_i - 1 + \gamma_i\lambda)/n_i)}{(\lambda/1) \cdot ((1 + \lambda)/2)\ldots((n - 1 + \lambda)/n)}$$

In the next sections we shall reformulate (11) and (13) for a positive inductive GC-system (Section 4) and a negative inductive GC-system (Section 5).

### 3.2 Generalized Special Values

Let $S$ be a non-empty subset of $W$ and let $s$ be the number of its elements. If, however, $s$ occurs as subscript it indicates $S$. $\bar{S}$ is the complement of $S: W - S$. Let $\gamma_s$ be equal to the sum of the initial special values of the elements of $S$. From (2) and (10) it follows that

(14) $$p(S/e_n) = (n_s + \gamma_s\lambda)/(n + \lambda)$$

in which $n_s$ is of course the number of occurrences in $e_n$ of $Q_i$ belonging to $S$. We call $p(S/e_n)$ a generalized special value and we assume that (14) includes the generalized initial special value $p(S)$, which is equal to $\gamma_s$.

Let $e_n = Q_{i_1}\ldots Q_{i_k}\ldots Q_{i_n}$ and let $e_n' = Q_{i_1}'\ldots Q_{i_k}'\ldots Q_{i_n}'$. We say that $e_n$ and $e_n'$ are $S$-equal with respect to the $k$-th coordinate if $Q_{i_k}$ and $Q_{i_k}'$ belong both to $S$ or to $\bar{S}$. Let $_se_n$ be the set of those $e_n'$ for which $e_n'$ and $e_n$ are $S$-equal with respect to all coordinates.

From (14) we see that $p(S/e_n)$ does not depend on the way in which $n_s$ and $(n - n_s)$ are composed out of the $n_i$ and therefore we have that $p(S/_se_n) = p(S/e_n)$. Analogous to the way in which we have obtained (11) from (3), via (6), and (10) we can now derive from (3) and (14)

(15) $$p(_se_n) = \frac{\gamma_s\lambda \cdot (1 + \gamma_s\lambda)\ldots(n_s - 1 + \gamma_s\lambda)}{\lambda \cdot (1 + \lambda)\ldots}$$
$$\frac{((1 - \gamma_s)\lambda) \cdot (1 + (1 - \gamma_s)\lambda)\ldots(n - n_s - 1 + (1 - \gamma_s)\lambda)}{\ldots(n - 1 + \lambda)}$$

in which the first part of the numerator vanishes if $n_s = 0$ and the second part if $n_s = n$.

From (15) we can easily see that there is in a GC-system also order indifference with respect to $S$: if $e_n$ and $e'_n$ are such that $n_s = n'_s$ then $p(_se_n) = p(_se'_n)$. Let $_s\ddot{e}_n$ be the union of the $\binom{n}{n_s}$ sets $_se'_n$ of those $e'_n$ for which $n_s = n'_s$ then we obtain, analogous to (12), from (2)

(16)     $$p(_s\ddot{e}_n) = \binom{n}{n_s} \cdot p(_se_n)$$

and therefore we have in combination with (15)

(17)     $$p(_s\ddot{e}_n) = \frac{(\gamma_s\lambda/1) \cdot ((1 + \gamma_s\lambda)/2) \dots ((n_s - 1 + \gamma_s\lambda)/n_s)}{(\lambda/1) \cdot ((1 + \lambda)/2) \dots}$$

$$\cdot \; \frac{((1 - \gamma_s)\lambda/1) \cdot ((1 + (1 - \gamma_s)\lambda)/2) \dots ((n - n_s - 1 + (1 - \gamma_s)\lambda)/(n - n_s)}{\dots ((n - 1 + \lambda)/n)} \; .$$

3.3. *First Interpretation of GC-systems: the urn-model* ($w < \infty$)

Let an urn contain $M(<\infty)$ coloured balls. Let $W$ be the set of different colours that occur in the urn. A ball with colour $Q_i$ will be called a $Q_i$-ball. Let $M_i$ be the (positive) number of $Q_i$-balls in the urn.

Let $\Delta$ be a finite integer. A trial is a random selection of one ball; if this selected ball appears to be a $Q_i$-ball we say that the trial has resulted in $Q_i$. If $\Delta$ is positive we replace the drawn ball and we add, moreover, $\Delta$ new balls to the urn with the same colour as the colour of the ball that has been drawn (and replaced). If $\Delta$ is negative we replace also the drawn ball but then we take out $-\Delta$ balls of the colour that has been drawn, if this is still possible. If $\Delta = 0$ we only replace the drawn ball. In summary we may say that a trial is a random selection of one ball and that it is followed, if possible, by the replacement of that ball and the 'addition' of $\Delta$ balls of the same colour. One might also say that the trials are random selections 'with replacement of $\Delta + 1$ balls'.

We are of course interested in sequences of the trials described. In case $\Delta$ is negative it may be impossible to add $\Delta$ balls in which case the process of successive trials is stopped. Suppose that it has been possible to perform $n$ trials. Let $e_n$ indicate the ordered sequence of results and let $n_i$ indicate the number of $Q_i$ among them. If the $n$-th trial resulted in $Q_j$ then it is possible to 'add' $\Delta$ $Q_j$-balls if and only if

(18)     $n_j \Delta + M_j \geq 0$.

Note that (18) is always fulfilled if $\Delta$ is non-negative.

Suppose now that (18) is satisfied and hence that the $(n + 1)$-th trial can be performed. We indicate the objective probability that this trial will result in $Q_i$ by $P(Q_i/e_n)$ and this probability is easy to calculate because all balls in the urn are supposed to be equally probable to be drawn. The number of balls in the urn is of course equal to $n\Delta + M$ and the number of $Q_i$-balls is equal to $n_i\Delta + M_i$. Hence we have

(19)     $P(Q_i/e_n) = (n_i\Delta + M_i)/(n\Delta + M)$.

For $n = n_i = 0$ (19) is in accordance with the initial objective probability: $P(Q_i) = M_i/M$. Now it is easily seen that if we replace $M_i/M$ by $\gamma_i$ and $M/\Delta$ by $\lambda$ that (19) transforms into the right side of (10), i.e., the special value of a GC-system. Of course it follows therefore also that the objective probability $P(e_n)$ of $e_n$ is equal to the right side of (11).

If $\Delta$ is non-negative, (19) gives always non-negative values for $P(Q_i/e_n)$ and the process can never be forced to stop.

If $\Delta$ is negative, then the largest number $N$ such that $P(Q_i/e_n)$ is non-negative, according to (19), for all $n_i = 0, 1, \ldots, n < N$ can be shown to be the largest $N$ for which

(20)     $N - 1 \leq M_{\min}/|\Delta|$

where $M_{\min}$ indicates the smallest $M_i$. The argument is essentially the same as the one used at the end of the Appendix to prove the $(-)$-version of the parameter-condition (9) for GC-systems and (20) turns into that condition after the relevant substitutions.

However, this number $N$ may be smaller than the smallest number of trials $N'$ after which the process may be forced to stop. To determine $N'$ we consider first the case that all $M_i$ are multiples of $|\Delta|$, i.e., let there be $N_i$ such that $M_i = N_i|\Delta|$. Of course, $M$ is now also a multiple of $|\Delta|$. If $e_n$ is such that, for some $i$, $n_i = N_i$ then the urn does not contain $Q_i$-balls anymore; hence, $P(Q_i/e_n) = 0$ such that following trials cannot result in $Q_i$. But as long as this does not hold for all $i$ the trials can be continued. The process is therefore forced to stop precisely at the moment the urn is exhausted: hence $N' = M/|\Delta|$ (which is an integer).

Consider now the case that not all $M_i$ are multiples of $|\Delta|$; let $M_b$ indicate the smallest $M_i$ that is not a multiple. From (18) we may conclude now, assuming the most unfavourable case, that $N'$ is the largest integer for which

(21)      $N' - 1 \leq M_b/|\Delta|.$

If $M_b = M_{\min}$, which is for instance the case if none of the $M_i$ is a multiple of $|\Delta|$, then (21) is equivalent to (20). In other words, only in this case is the smallest number of trials after which the process may be forced to stop is not larger than but equal to the largest number for which $P(Q_i/e_n)$ cannot be negative according to (19).

In conclusion we may say that the described urn model gives rise to a system of objective probabilities which is mathematically equivalent to a GC-system. It is also clear that any GC-system can be interpreted in terms of this urn model as soon as $\lambda$ and all $\gamma_i$ are rational numbers and $w$ is finite. Note that in the case $\lambda = \pm\infty$ this interpretation reduces to the well-known urn model of successive random sampling with replacement and in the case $\lambda = -M$ to random sampling without replacement.*

The urn model is a generalization of the model introduced by Polya [30], which is restricted to the case $w = 2$. Friedman has generalized Polya's urn model [12] to trials which are not only succeeded by replacement of a constant number of balls with the last drawn colour but also with the addition of a constant number of balls with the opposite colour. This model can of course be generalized to an arbitrary number of colours in terms of addition of a constant number of balls for each of the other colours. It is easy to see that in this model the principle of order indifference is no longer fulfilled and it is therefore outside the scope of this study.

---

* After reading the manuscript, Prof. R. Jeffrey suggested in a personal letter the following simple formulation of urn-models for GC-systems in case the $\gamma_i$ are rational numbers and $\lambda$ is a, possibly infinite, integer such that the $\gamma_i\lambda$ values are also integers. For each value/interval of $\lambda$ we fix the numbers $M$ and $M_i$ in terms of $\lambda$ and $\gamma_i$. It is then easily shown, on the basis of (19) and (10), that $\Delta$ assumes the mentioned definite values.

$\lambda = \infty$     $M$ arbitrarily such that $M_i = {}_{df}\gamma_i M$ gives integers
            $\Delta = 0$: random sampling with (simple) replacement.
$0 < \lambda < \infty$   $M = \lambda$, $M_i = \lambda_i M$
            $\Delta = 1$: random sampling 'with double replacement'.
$\lambda = 0$       $M$ arbitrarily such that $M_i = {}_{df}\gamma_i M$ gives integers
            $\Delta = \infty$: random sampling 'with infinite replacement'.
$\lambda < 0$       $M = -\lambda = {}_{df}\eta$, $M_i = \gamma_i M$
            $\Delta = -1$: random sampling without replacement.

## 3.4. *Mathematical Expectations According to GC-systems*

In this subsection we shall calculate the most interesting mathematical expectations of the random variables to which a GC-system gives rise. We shall first introduce the standard definitions and symbolizations (see, e.g., Feller [10], Chapter 9).

Let $X$ be a countable set of real numbers and let $\mathbf{x}$ be a random variable with respect to $X$: there is a probability function $P$ such that, for $x$ in $X$, $P(\mathbf{x} = x)$ represents the probability that $\mathbf{x}$ assumes the value $x$. In case of two random variables on $X$ we suppose to have the joint distribution $P(\mathbf{x}_1 = x_1, \mathbf{x}_2 = x_2)$. Let $F(\mathbf{x})$ be a real-valued function on $X$. The expectation $E(F(\mathbf{x}))$ of $F(\mathbf{x})$ is now defined as $\sum_{x \in X} F(x)P(\mathbf{x} = x)$. It is easy to prove that $E(aF(\mathbf{x}) + b)$, in which $a$ and $b$ are constants, is equal to $aE(F(\mathbf{x})) + b$. In the case of two random variables the definition of expectation and the last-mentioned property are analoguous.

The *mean* of $\mathbf{x}$ is simply defined as $E(\mathbf{x})$. The *variance* $\sigma^2(\mathbf{x})$ of $\mathbf{x}$ is defined as $E((\mathbf{x} - E(\mathbf{x}))^2)$, which can easily be proved to be equal to $E(\mathbf{x}^2) - (E(\mathbf{x}))^2$. The *covariance* cov $(\mathbf{x}_1, \mathbf{x}_2)$ of $\mathbf{x}_1$ and $\mathbf{x}_2$ is defined as $E((\mathbf{x}_1 - E(\mathbf{x}_1))(\mathbf{x}_2 - E(\mathbf{x}_2)))$ which can be proved to be equal to $E(\mathbf{x}_1\mathbf{x}_2) - E(\mathbf{x}_1)E(\mathbf{x}_2)$. Finally the *correlation coefficient* $\rho(\mathbf{x}_1, \mathbf{x}_2)$ between $\mathbf{x}_1$ and $\mathbf{x}_2$ is defined as cov $(\mathbf{x}_1, \mathbf{x}_2)/(\sigma(\mathbf{x}_1)\sigma(\mathbf{x}_2))$.

Let $\mathbf{i}_n$ be the random variable assuming the value 1 if the $n$-th coordinate of $e_n \in W^n$ (or, in terms of the urn model, the $n$-th trial) is occupied by (results in) $Q_i$ and the value 0 otherwise. We define the random variable $\mathbf{n}_i$ as the sum of the random variables $\mathbf{i}_1, \mathbf{i}_2, \ldots, \mathbf{i}_n$.

A GC-system gives rise to a probability distribution on $\mathbf{i}_n$ in an obvious way, and we shall indicate that distribution also by $p$. We have of course $p(\mathbf{i}_n = 1) = p(W^{n-1}Q_i)$ and from the principle of order indifference it follows immediately that $p(\mathbf{i}_n) = p(\mathbf{i}_1)$ and the last value is equal to $\gamma_i$. On the basis of the same principle we obtain that the joint distribution $p(\mathbf{i}_n = 1/0, \mathbf{j}_{n'} = 1/0)$, $n' \neq n$, is equal to $p(\mathbf{i}_1 = 1/0, \mathbf{j}_2 = 1/0)$ and it is easy to calculate the values for all eight combinations (four in case $i \neq j$ and four in case $i = j$). Finally, the joint distribution $p(\mathbf{i}_n = 1/0), \mathbf{j}_n = 1/0)$, $i \neq j$, reduces immediately to the value 0 for the combination $(1, 1)$, $\gamma_i$ for $(1, 0)$, $\gamma_j$ for $(0, 1)$ and $1 - \gamma_i - \gamma_j$ for $(0, 0)$.

In the following survey we shall give for each random variable its mean and variance. With respect to two random variables we shall start with the expectation of their product, followed by their covariance and their correlation coefficient. Proofs will be given only if they are not too easy. We start with

(22.1)    $E(\mathbf{i}_n) = \gamma_i$            (22.2)    $\sigma^2(\mathbf{i}_n) = \gamma_i(1 - \gamma_i).$

For $n \neq n'$ we have

(23.1)    $E(\mathbf{i}_n\mathbf{i}_{n'}) = \gamma_i(1 + \gamma_i\lambda)/(1 + \lambda)$

(23.2)    $\text{cov}(\mathbf{i}_n, \mathbf{i}_{n'}) = \gamma_i(1 - \gamma_i)/(1 + \lambda)$

(23.3)    $\rho(\mathbf{i}_n, \mathbf{i}_{n'}) = 1/(1 + \lambda).$

For $i \neq j$ we have

(24.1)    $E(\mathbf{i}_n\mathbf{j}_n) = 0$

(24.2)    $\text{cov}(\mathbf{i}_n, \mathbf{j}_n) = -\gamma_i\gamma_j$

(24.3)    $\rho(\mathbf{i}_n, \mathbf{j}_n) = -\sqrt{\gamma_i\gamma_j/((1 - \gamma_i)(1 - \gamma_j))}.$

For $n \neq n'$ and $i \neq j$ we have

(25.1)    $E(\mathbf{i}_n\mathbf{j}_{n'}) = \gamma_i\gamma_j\lambda/(1 + \lambda)$

(25.2)    $\text{cov}(\mathbf{i}_n, \mathbf{j}_{n'}) = -\gamma_i\gamma_j/(1 + \lambda)$

(25.3)    $\rho(\mathbf{i}_n, \mathbf{j}_{n'}) = -\sqrt{\gamma_i\gamma_j/((1 - \gamma_i)(1 - \gamma_j))}/(1 + \lambda).$

Note that it follows from (23.3), (24.3) and (25.3) that

(26)      $\rho(\mathbf{i}_n, \mathbf{j}_{n'}) = \rho(\mathbf{i}_n, \mathbf{i}_{n'})\rho(\mathbf{i}_n, \mathbf{j}_n),\quad n \neq n'$ and $i \neq j.$

With respect to $\mathbf{n}_i$ we get

(27.1)    $E(\mathbf{n}_i) = n\gamma_i$

(27.2)    $\sigma^2(\mathbf{n}_i) = n\gamma_i(1 - \gamma_i)(n + \lambda)/(1 + \lambda).$

For the proof of (27.2) we need the intermediate stage

(28)      $\sigma^2(\mathbf{n}_i) = n\sigma^2(\mathbf{i}_k) + n(n - 1)\,\text{cov}(\mathbf{i}_k, \mathbf{i}_{k'}),\ k \neq k'.$

For $i \neq j$ (and fixed $n$) we have

(29.1)    $E(\mathbf{n}_i\mathbf{n}_j) = n(n - 1)\gamma_i\gamma_j\lambda/(1 + \lambda)$

(29.2)    $\text{cov}(\mathbf{n}_i, \mathbf{n}_j) = -n\gamma_i\gamma_j(n + \lambda)/(1 + \lambda)$

(29.3)    $\rho(\mathbf{n}_i, \mathbf{n}_j) = -\sqrt{\gamma_i\gamma_j/((1 - \gamma_i)(1 - \gamma_j))} = \rho(\mathbf{i}_n, \mathbf{j}_n).$

The proof of (29.1) has as intermediate stage

(30)      $E(\mathbf{n}_i\mathbf{n}_j) = nE(\mathbf{i}_k\mathbf{j}_k) + n(n - 1)E(\mathbf{i}_k\mathbf{j}_{k'}),\quad k \neq k'.$

Of course we may construe random variables analoguous to $i_n$ and $n_i$ under the condition that the first $k$ trials have resulted in $e_k$. For the calculation of the corresponding conditional expectations we have only to substitute in (22)–(30) the conditional special values for the initial special values.

Two random variables are called uncorrelated if their covariance is 0. From (23.2) and (25.2) it follows that $i_n$ and $j_{n'}$ ($n \neq n'$) are uncorrelated if $\lambda = \pm\infty$. This is not surprising for it is clear that they are even independent in this case. For these two values of $\lambda$ we have that $\sigma^2(n_i) = n\sigma^2(i_n)$. We define now the square of the *dispersion coefficient* $D(n_i)$ of a GC-system as follows: $D^2(n_i) = {}_{df}\sigma^2(n_i)/\sigma^2_{\lambda=\infty}(n_i)$ and it is easy to see that

$$(31) \qquad D^2(n_i) = (n + \lambda)/(1 + \lambda) = n/(1 + \lambda) + \lambda/(1 + \lambda).$$

From (31) it follows that the dispersion coefficient does only depend on $n$ (and $\lambda$) but not on $i$. Therefore we may indicate its square also by $D^2(n)$. For positive finite values of $\lambda$, $D^2(n)$ lies between 1 and $n$; for $\lambda = 0$ it is equal to $n$; for negative finite values it is smaller than 1; of course it is equal to 1 for the two infinite values of $\lambda$.

3.5. *Non-inductive* ($\lambda = \pm\infty$) *and Extreme-inductive* ($\lambda = 0$) *GC-systems*
$\lambda = +\infty$. The values $+\infty$ for $\lambda$ in a GC-system correspond to the value 1 for $\alpha$. For this value of $\alpha$ the parameter-condition (9) is satisfied and we have already seen that (10) reduces to

$$(32) \qquad p(Q_i) = p(Q_i/e_n) = \gamma_i$$

and we may conclude that the special values do not change. The urn-model interpretation for this particular GC-system corresponds to successive random sampling with replacement and this is a well-known and well-described process in applied probability theory.

If we apply this GC-system it means that we treat the coordinates as independent (experiments). In other words the results of previous experiments do not play a role: we do not use experience. Now the use of experience can be seen as a rough characteristic of inductive procedures and therefore we call a GC-system in which $\lambda = \pm\infty (\alpha = 1)$ *non-inductive*.

In a non-inductive GC-system (11) reduces to

$$(33) \qquad p(e_n) = \Pi(\gamma_i)^{n_i}$$

and we have already seen that its dispersion coefficient equals 1 for all $n$: there is no dispersion.

$\lambda = 0$. The value 0 for $\lambda$ in a GC-system corresponds to the value 0 for $\alpha$. Again the parameter-condition is satisfied. Now (10) reduces to

(34.1)    $p(Q_i) = \gamma_i$        (34.2)   $p(Q_i/e_n) = n_i/n$.

The latter implies that the conditional special values are equal to the relative frequencies according to $e_n$. In [3], Carnap calls this system with $\gamma_i = 1/w$ the straight rule. For the present generalized case we speak of a *straight* or *extreme-inductive* GC-system.

The extreme-inductive system is basic for certain statistical procedures. The relative frequency of a property in a, randomly drawn, sample of a population is frequently used as an estimate of the relative frequency in the whole population, provided that the sample is large enough in comparison with the size of the population. This estimate can then be used as special value in a non-inductive GC-system for the calculation of the probability that a new sample will be such and such. However, the assumption that the sample is large enough is essential, for if this is not the case the described procedure is disastrous. Suppose for instance that the original sample contains only one element. The procedure assigns then the value 1 to the hypothesis that all individuals of a new sample will have the same property as that of the individual of the original sample. To be fair, in this example we are not yet in a position to apply a proper non-inductive GC-system, with the $\gamma_i$'s equal to the respective relative frequencies, for the principle of initial possibility is only satisfied as soon as the original sample has positive relative frequencies for all properties.

In an extreme-inductive GC-system we have

(35.1)    $p(e_n) = \gamma_i$   for $e_n$ and $i$ such that $n_i = n$

(35.2)    $p(e_n) = 0$   if $n_i = 1, 2, \ldots, n - 1$ for some $i$.

In other words if $e_n$ is homogeneous in, say, $Q_i$ it has (positive) probability $\gamma_i$, if it is non-homogeneous it has zero probability.

3.6. *Carnapian Systems (C-systems)*
The principle of initial possibility (PIP) states that all initial special values are positive. In case $w$ is finite we obtain a very interesting subclass of GC-systems if we replace this principle by the stronger one:

(PIE)    *principle of initial equipossibility*
         $\gamma_i = 1/w$.

If we should have started with this principle instead of **PIP** then we would have been able to prove T2 for the case that all $\gamma_i$ as well as $\gamma_{\min}$ are equal to $1/w$. We would also have obtained this result if we had replaced **PRR** by the stronger principle to the effect that $p(Q_i/e_n)$ depends only on $i$ via $n_i$ (i.e., $p(Q_i/e_n) = f(n, n_i)$).

We introduce now the following:

DEFINITION: a Carnapian system (C-system) is a consistent probability pattern w.r.t. $W$, $W^2$, $W^3$, . . . ., $(W^N)$ of a finite set $W$, for which there is a real number $\lambda$ such that

(36)    $p(Q_i/e_n) = f(n, n_i) = (n_i + \lambda/w)/(n + \lambda)$
holds for all $n < N$ and all $i$.

It is easy to see that the parameter-condition now reads

(37)    $0 \leq \alpha = \lambda/(1 + \lambda) \leq (N - 1)/((N - 1) - 1/w)$

and that this condition is included in the definition. For negative values of $\lambda$ (37) reduces to: $-\lambda = \eta \leq w(N - 1)$ and there are only such values if $N$ is also finite.

As far as non-negative values of $\lambda$ are concerned a C-system corresponds exactly to a member of Carnap's continuum of inductive methods [3]. As we have said in the Introduction of this chapter Carnap had a particular application in mind with his continuum and it seems now appropriate to sketch this application.

Carnap starts with a language system with $N$ individual constants $a_1, a_2, \ldots, a_N$ and a finite number of primitive predicates. These primitive predicates give rise to the so-called $Q$-predicates: (in interpreted language) a $Q$-property specifies a sub-class of primitive porperties such that an individual has this $Q$-property if and only if it has all the primitive properties of the subclass and none of the remaining primitive properties. The number of $Q$-predicates is of course equal to $2^v$, if $v$ is the number of primitive predicates. In this interpretation $e_n$ is an ordered description of $n$ individuals in terms of the $Q$-predicates. A special value has now to be considered as the 'logical probability' of the hypothesis that, again in interpreted language, an individual not included in the set of individuals described by $e_n$ has a particular $Q$-property. Of course the continuum gives rise to values for all different descriptions of the $N$ individuals in terms of the $Q$-properties, the so-called state-descriptions $e_N$. From

Carnap's logical point of view it followed that he had to take equal
initial special values for all $Q$-predicates (which may of course be seen as
an application of our so-called second-order principle of indifference
MPI). Carnap's restriction to non-negative values of $\lambda$ was based on his
assumption that the special values should be non-decreasing functions of
the relative frequency. Stegmüller has reformulated and generalized
Carnap's intended application in model-theoretical terms [34].

### 4. ANALYSIS OF POSITIVE INDUCTIVE GC-SYSTEMS ($0 < \lambda < \infty$)

#### 4.1. *Possible Reformulations*
In this section we shall study an arbitrary GC-system with a positive
finite (real) value for $\lambda$: we have called such a GC-system positive inductive.
With the use of the abbreviation

(38.1)    $\pi(k, x) = x(x + 1) . (x + 2) \ldots (x + k - 1), \quad k = 1, 2, 3, \ldots$

(38.2)    $\pi(0, x) = 1$

we may reformulate (11) as follows:

(39)    $p(e_n) = \{\Pi \pi(n_i, \gamma_i \lambda)\} / \pi(n, \lambda).$

The so-called gamma-function

(40)    $\Gamma(x) = \int_0^\infty e^{-t} t^{x-1} dt, \quad x > 0$

has the well-known and easy provable properties

(41.1)    $\Gamma(x + 1) = x \Gamma(x)$          (41.2)    $\Gamma(n + 1) = n!$

and this enables us to rewrite (39) as

(42)    $p(e_n) = \dfrac{\Pi\{\Gamma(n_i + \gamma_i \lambda) / \Gamma(\gamma_i \lambda)\}}{\Gamma(n + \lambda) / \Gamma(\lambda)}.$

If $1 \neq w < \infty$ we may reformulate (42) in terms of the so-called
(generalized) beta-function $B(x_1, \ldots, x_w)$, or simply $B(x_i)$:

(43)    $\int_{B_{w-1}} t_2^{x_2-1} \ldots t_w^{x_w-1} (1 - (t_2 + \ldots t_w))^{x_1-1} dt_2 \ldots dt_w$

in which all $x_i$ are supposed to be positive and $B_{w-1}$ is the set of all $(w - 1)$-tuples of non-negative numbers for which the sum does not exceed 1.

We need for our reformulation the following general theorem:

$$(44) \qquad B(x_i) = \{\Pi \Gamma(x_i)\}/\Gamma(\Sigma x_i), \quad x_i > 0.$$

For $w = 2$ this theorem is well-known in (mathematical) analysis, its proof is not elementary. The inductive step of the proof of the generalization to (44) goes as follows: suppose (44) has been proved for $w - 1$, then we integrate in (43) with respect to $t_w$ for fixed $t_2, \ldots, t_{w-1}$, and substitute $t_w = v(1 - t_2 \ldots - t_{w-1})$; then $v$ runs from 0 to 1 and $t_2, \ldots, t_{w-1}$ over $B_{w-2}$; with the use of the inductive hypothesis and the theorem for $w = 2$ the proof can easily be completed. For the case $w = 1$ we define $B(x) = 1$ for $x > 0$, in accordance with (44).

Now, with the use of (44), (42) transforms into

$$(45) \qquad p(e_n) = B(n, + \gamma_i\lambda)/B(\gamma_i\lambda), \quad w < \infty.$$

From (12), (41) and (42) we can get

$$(46) \qquad p(\ddot{e}_n) = \frac{\Pi\{\Gamma(n_i + \gamma_i\lambda)/(\Gamma(\gamma_i\lambda)\Gamma(n_i + 1))\}}{\Gamma(n + \lambda)/(\Gamma(\lambda)\Gamma(n + 1))}$$

and from (12), (41) and (45)

$$(47) \qquad p(\ddot{e}_n) = \frac{n!}{(n + w - 1)!} \cdot \frac{B(n_i + \gamma_i\lambda)}{B(\gamma_i\lambda)B(n_i + 1)}, w < \infty.$$

For the generalized special values we can prove theorems analogous to (45) and (47) starting from (15) and (16). We get, unconditionally,

$$(48) \qquad p(_s e_n) = \frac{B(n_s + \gamma_s\lambda, n - n_s + (1 - \gamma_s)\lambda)}{B(\gamma_s\lambda, (1 - \gamma_s)\lambda)}$$

$$(49) \qquad p(_s\ddot{e}_n) = \frac{1}{n + 1} \frac{B(n_s + \gamma_s\lambda, n - n_s + (1 - \gamma_s)\lambda)}{B(\gamma_s\lambda, (1 - \gamma_s)\lambda)B(n_s + 1, n - n_s + 1)}$$

in which $B(x_i)$ has been replaced by $B(x_1, x_2, \ldots)$. These two theorems are both based on the following direct reformulation of (15) in terms of (38):

$$(50) \qquad p(_s e_n) = \frac{\pi(n_s, \gamma_s\lambda)\pi(n - n_s, (1 - \gamma_s)\lambda)}{\pi(n, \lambda)}.$$

4.2. *Generalized Special Values as Weighted Means*
If we rewrite (41) as follows

$$(51) \qquad p(S/e_n) = \frac{n}{n + \lambda} \cdot \frac{n_s}{n} + \frac{\lambda}{n + \lambda} \cdot \gamma_s$$

then we see that the conditional generalized special value is the weighted mean of $n_s/n$ and $\gamma_s$ if $\lambda$ is non-negative for

$$(52) \qquad \frac{n}{n + \lambda} \geq 0, \quad \frac{\lambda}{n + \lambda} \geq 0; \quad \frac{n}{n + \lambda} + \frac{\lambda}{n + \lambda} = 1.$$

For positive values of $\lambda$ these weights are also defined if $n = 0$. For positive finite values of $\lambda$ the weights depend on $n$ in a very interesting way. The weight $\lambda/(n + \lambda)$ of $\gamma_s$ decreases with $n$ from 1 to 0 and, conversely, the weight $n/(n + \lambda)$ of $n_s/n$ increases with $n$ from 0 to 1. In other words, the larger $n$, the larger the weight of the relative frequency of $S$ in $e_n$ and the smaller the weight of the initial special value of $S$.

If $\lambda$ is infinite the weight of the initial special value is uniformly equal to 1 and the weight of the relative frequency is uniformly equal to 0. On the other hand, for $\lambda = 0$ and $n > 0$ the weight of the relative frequency is 1 and the weight of the initial special value is 0. In the light of these observations the terminology introduced in Section 3.5 (non-inductive/extreme-inductive GC-systems) gets a precise base. Both special cases of GC-systems may be said to be improperly inductive if we mean by 'inductive' that the weights do change.

In view of the described behaviour of the weights in case $\lambda$ is a positive finite number it is in agreement with this terminology that we have called such a GC-system inductive. We talk in this case, moreover, about a *positive* inductive GC-system because of the following property

$$(53) \qquad p(S/e_n) > p(S/e_n') \quad \text{if } n_s > n_s'$$

that is, the larger the number of occurrences in $e_n$ the larger the (generalized) special value. Note that (53) does also hold, trivially, in an extreme-inductive GC-system. In agreement with the literature we may say that (53) represents the *property* (or principle) *of positive relevance* (see, e.g., Stegmüller |34|, p. 474) and that a non-inductive GC-system has the property of zero relevance. As we shall see in the next section a negative inductive GC-system has the property of negative relevance (the larger the number of occurrences the smaller $p(S/e_n)$).

Suppose we apply a positive inductive GC-system with infinite $N$ in a case in which the underlying interpretation guarantees that $n_s/s$ converges, with certainty, to a certain limit. Then it is easy to see from (51) that $p(S/e_n)$ also converges to this limit. (A particular class of such underlying interpretations are infinite sequences of mutually independent experiments with constant probabilities: the so-called multinomial contexts, to be studied in Chapter 7.) It also follows directly from (51) that the speed of convergence decreases by increasing $\lambda$.

We shall finish this subsection by a theorem with far-reaching consequences. Though the $p$-value of the infinite product $W^\infty$ is, of course, equal to 1, we have

(T3)     In a positive inductive GC-system
$$p(S^\infty) = 0 \text{ for } S \subset W, \quad \varnothing \neq S \neq W.$$

*Proof:* Of course we have that $p(S^\infty) = \lim\limits_{N \to \infty} p(S^N)$ and it follows from (15) that

(*)     $$p(S^N) = \prod_{n=0}^{N-1} \{(n + \gamma_s\lambda)/(n + \lambda)\} = \prod_{n=0}^{N-1} \{1 - (1 - \gamma_s)\lambda/(n + \lambda)\}$$

There is a well-known theorem that (*) tends to zero for $N \to \infty$ if and only if $\Sigma\{(1 - \gamma_s)\lambda/(n + \lambda)\} = \infty$, which is true for $0 < \gamma_s < 1$ and $0 < \lambda < \infty$, q.e.d.

It is easy to check that the theorem holds also for $\lambda = \infty$, but that $p(S^\infty) = \gamma_s$ for $\lambda = 0$ and that in this case $p(S^\infty/e_n)$ is 1 if $n_s = n$ and 0 otherwise. Finally, if $\lambda$ is negative $p(S^\infty)$ is not defined.

It is important to see that for $0 < \lambda \leq \infty$, $p(S^\infty/e_n)$ remains zero for all $n$, even if $n_s = n$, in spite of the fact that $p(S/e_n)$ goes to 1 if $n = n_s$ goes to $\infty$. The point is of course that the latter is smaller than 1 for all finite $n$. This implies that we keep expecting that the coordinates larger than $n$ will be non-homogeneous with respect to $S$ and if $N$ is infinite this will be the case for infinitely many coordinates. In certain underlying interpretations of GC-systems $S^\infty$ corresponds to a general or universal statement with respect to a denumerable universe. From our theorem it follows that, if we should apply a GC-system directly in such cases, the general statements would always have the $p$-value zero.

4.3. *Second Interpretation of GC-systems: Repeated Experiments Governed by a Density-function* $(w < \infty)$

Let $w$ be finite and let $\lambda$ be some positive finite number. Suppose for a moment that $W$ is the set of elementary outcomes of an experiment and that $p_i$ is the probability of the outcome $Q_i$. For a sequence of $n$ of these experiments we have that, if they are independent, the probability of the sequence of elementary outcomes $e_n$ is equal to $\Pi p_i{}^{n_i}$. Now let us assume that the $p_i$ are not fixed but that $p_2, p_3, \ldots, p_w$ (and therefore $p_1$, by $p_1 = 1 - (p_2 + \ldots + p_w)$) are stochastically determined by a volume-density function $q(t_2 \ldots t_w)$ on $B_{w-1}$ (defined in Subsection 4.1). Such a density function satisfies:

$$(54.1) \quad 0 \leq q(t_2 \ldots t_w) \qquad (54.2) \quad \int_{B_{w-1}} q(t_2 \ldots t_w) dt_2 \ldots dt_w = 1$$

and the probability $P(e_n)$ of $e_n$ is given by

$$(55) \qquad P(e_n) = \int_{B_{w-1}} q(t_2 \ldots t_w)\{\Pi t_i^{n_i}\} dt_2 \ldots dt_w,$$

$$t_1 = 1 - t_2 - \ldots - t_w.$$

It is easy to see that the particular density function

$$(56) \qquad q(t_2 \ldots t_w) = \{\Pi t_i^{\gamma_i \lambda - 1}\}/B(\gamma_i \lambda), \quad t_1 = 1 - t_2 - \ldots - t_w$$

satisfies (54) because of (43) and the definition of $B_{w-1}$. If we substitute (56) in (55) we see from (43) and (45) that $P(e_n) = p(e_n)$ and therefore also that $P(Q_i/e_n) = p(Q_i/e_n)$.

In conclusion we may say that a positive inductive GC-system with finite $w$ may be interpreted as a sequence of repeated experiments for which the probabilities of the elementary outcomes are governed by the density function (56).

If the $p_i$ were fixed we would of course have that $n_i/n$ converges with certainty to $p_i$ by the strong law of large numbers. In the same way we have in the described situation that $n_j/n$ converges to the probability of $Q_j$; however, this probability is not fixed but stochastically determined by the marginal (probability) density $q_j(t)$ of $p_j$, given by

$$(57) \qquad q_j(t) = t^{\gamma_j \lambda - 1}(1 - t)^{(1 - \gamma_j)\lambda - 1}/B(\gamma_j \lambda, (1 - \gamma_j)\lambda).$$

This may be proved, for $j = 2$, by descending induction: if $p_2, \ldots, p_w$ have joint density (56), then $p_2, \ldots, p_{w-1}$ have a similar density on $B_{w-2}$ with $\gamma_1, \gamma_2, \ldots, \gamma_{w-1}$ replaced by $\gamma_1 + \gamma_w, \gamma_2, \ldots, \gamma_{w-1}$. This is found by integrating (56) with respect to $t_w$, the interval of integration being $(0, 1 - t_2 - \ldots - t_{w-1})$. For $j \neq 2$ we obtain the result by changing the indices $1, 2, \ldots, w$.

From (10) it is seen that the conditional probability measure given $e_n$ on the Cartesian product $WWW \ldots$ is again a GC-system, with parameters $\gamma_i'$ and $\lambda'$ defined by

(58)    $\gamma_i' = (n_i + \gamma_i \lambda)/(n + \lambda), \quad \lambda' = n + \lambda.$

This conditional probability measure assigns, by definition, probability $p'(e_k) = p(e_n e_k/e_n)$, where $e_k \in W^k$.

By what has been proved above, this conditional probability measure admits the interpretation of experiments with stochastic probabilities $p_1, \ldots, p_w$, the density of $p_2, \ldots, p_w$ on $B_{w-1}$ given by

(59)    $q(t_2 \ldots t_w/e_n) = \{\Pi t_i{}^{\gamma_i' \lambda' - 1}\}/\mathrm{B}(\gamma_i' \lambda'), \quad t_1 = (1 - t_2 - \ldots - t_w).$

From (58), (56) and (45) it is seen now that

(60)    $q(t_2 \ldots t_w/e_n) = q(t_2 \ldots t_w) \cdot \{\Pi t_i^{n_i}\}/P(e_n),$

$t_1 = (1 - t_2 - \ldots - t_w)$

which is, as expected, the conditional probability density given $e_n$ of the original $p_2, \ldots, p_w$.

To reformulate this result, the interpretation of a GC-system as a sequence of repeated experiments governed by a density function remains valid after the performance of a number of experiments. The density function changes however from (56) to (59).

It is easy to check that the density functions that have been introduced are only defined for positive finite values of $\lambda$ and finite values of $w$.

### 4.4. *Principle of Structural Indifference* ($w < \infty$): $C^*$-*systems* ($\lambda = w$)

In Subsection 3.6 we have studied some aspects of what happens if the principle of initial possibility (PIP) is replaced by the stronger principle of initial equipossibility (PIE). This replacement was only possible for the case that $w$ is finite. In this subsection we shall study what happens if we replace PIP by a principle which is still stronger than PIE:

(PSI)    *principle of structural indifference*
$p(\ddot{e}_n) = p(\ddot{e}_n').$

This principle coincides with PIE for $n = 1$ and it presupposes therefore also that $w$ is finite. If we assume PSI then we have of course that $p(\ddot{e}_n)$ is the inverse of the number of different $\ddot{e}_n$. To calculate this number we have to realize that this number is equal to the number of different ways in which $n$ (indistinguishable) balls can be distributed in $w$ (different) cells and it is well known (Feller [10], p. 38) that this is equal to $\binom{n + w - 1}{w - 1}$.

We have therefore

(61)  $p^*(\ddot{e}_n) = n!(w - 1)!/(n + w - 1)!$

in which the star indicates that we have assumed PSI.

Now we add the strong version of the principle of indifference (SPOI), which permits us to use (12). From (61) and (12) follows

(62)  $p^*(e_n) = \{(w - 1)! \, \Pi n_i!\}/(n + w - 1)!$

and from this by (38)

(63)  $p^*(e_n) = \{\Pi \pi(n_i, 1)\}/\pi(n, w)$.

It is easy to check now that $p^*$ is a consistent probability pattern satisfying POI, PRR and PIE, that is, $p^*$ is a Carnapian system. By comparison of (63) and (39) we see immediately that it is a C-system with $\lambda = w$. We call such a system a $C^*$-system. For the (generalized) special values we obtain now from (63)

(64.1)  $p^*(Q_i/e_n) = (n_i + 1)/(n + w)$

(64.2)  $p^*(S/e_n) = (n_s + s)/(n + w)$

using (2) and (5).

From (64.1) we see that PSI and SPOI apparently imply the other principles that were needed to derive GC-systems: the principle of restricted relevance (PRR) and the special principle of linearity (SPL) for the case $w = 2$. PSI may of course be seen as an application of MPI, the second-order principle of indifference. PSI has appeared to be useful in certain fields of physics (see, e.g., Feller [10], p. 41) and constitutes so-called Bose–Einstein statistics.

There is an interesting illustration of the way in which MPI may be seen to be applied in $C^*$-systems if we look at the density interpretation of the preceding subsection. If we substitute $\lambda = w$ and $\gamma_i = 1/w$ in (56) we get

(65)  $q^*(t_2 \ldots t_w) = (w - 1)! = 1/\displaystyle\int_{B_{w-1}} dt_2 \ldots t_w = 1/\text{volume } (B_{w-1})$

and by substitution of this in (55) we obtain

$$(66) \qquad p^*(e_n) = \left. \int_{B_{w-1}^{\cdot}} (1 - t_2 \ldots - t_w)^{n_1} t_2^{n_2} \ldots t_w^{n_w} \, dt_2 \ldots dt_w \middle/ \right.$$

$$\text{volume}(B_{w-1})$$

in which we have used $p^* = P^*$. The interpretation of (66) is straightforward: the probabilities $t_2, \ldots, t_w$ are randomly chosen in $B_{w-1}$ or, equivalently, $t_1, t_2, \ldots, t_w$ are randomly chosen out of the set of $w$-tuples of non-negative numbers summing to unity. According to (57), in this particular case, the marginal probability density of $p_j$ is

$$(67) \qquad q_j^*(t) = (w - 1)(1 - t)^{w-2}.$$

## 5. ANALYSIS OF NEGATIVE INDUCTIVE GC-SYSTEMS ($\lambda < 0$)

### 5.1. *Possible Reformulations*
In this section we shall mainly concentrate our attention on an arbitrary GC-system with a negative finite (real) value for $\lambda$ which will be replaced by $-\eta$. We assume of course also that $N < \infty$ and $w < \infty$, and if not otherwise stated that the parameter-condition ($\eta \geq (N - 1)/\gamma_{\min}$) is satisfied. For the special values we repeat (10−):

$$(68) \qquad p(Q_i/e_n) = (\gamma_i \eta - n_i)/(\eta - n).$$

Substitution of $-\eta$ for $\lambda$ transforms (11) into

$$(69) \qquad p(e_n) = \frac{\prod_{i:n_i \neq 0} \{(\gamma_i \eta - n_i + 1) \ldots (\gamma_i \eta - 1)\gamma_i \eta\}}{(\eta - n + 1) \ldots (\eta - 1)\eta}$$

and this may be rewritten with the use of (38) as:

$$(70) \qquad p(e_n) = \frac{\Pi[\{\gamma_i \eta/(\gamma_i \eta - n_i)\}\pi(n_i, \gamma_i \eta - n_i)]}{(\eta/(\eta - n))\pi(n, \eta - n)}.$$

We can reformulate (69) in terms of the gamma-function (42) and the beta-function (44) as follows:

$$(71) \qquad p(e_n) = \frac{\Pi\{\Gamma(\gamma_i \eta + 1)/\Gamma(\gamma_i \eta - n_i + 1)\}}{\Gamma(\eta + 1)/\Gamma(\eta - n + 1)}$$

$$(72) \qquad p(e_n) = \frac{B(\gamma_i \eta + 1)\Gamma(\eta + w)\Gamma(\eta - n + 1)}{B(\gamma_i \eta - n_i + 1)\Gamma(\eta - n + w)\Gamma(\eta + 1)}.$$

Using the standard definition $1/\Gamma(0) = 0$, it is easy to check that these reformulations are always adequate, but that this would not be the case for $n = N$ if $\eta$ might assume the extreme value $N - 1$, which is however excluded by the parameter condition, $w < \infty$ and PIP.

Substitution of (70) in (12) leads to

$$(73) \qquad p(\ddot{e}_n) = \frac{---\left\{\begin{array}{cc} \gamma_i\eta & \Gamma(\gamma_i\eta) \\ - -\left\{ n_i(\gamma_i\eta - n_i) & \Gamma(n_i)\Gamma(\gamma_i\eta - n_i)\right\} \end{array}\right.}{\begin{array}{cc} \eta & \Gamma(\eta) \\ n(\eta - n) & \Gamma(n)\Gamma(\eta - n) \end{array}}$$

which is, in view of (44), equal to

$$(74) \qquad p(\ddot{e}_n) = \frac{B(\gamma_i\eta)}{B(n_i)B(\gamma_i\eta - n_i)} \frac{\prod\{\gamma_i\eta/(n_i(\gamma_i\eta - n_i))\}}{\eta/(n(\eta - n))}.$$

For the generalized special values we obtain from (15) by replacing $\lambda$ by $-\eta$ and using (38):

$$(75) \qquad p(_se_n) = \frac{\gamma_s\eta}{(\gamma_s\eta - n_s)} \frac{(1 - \gamma_s)\eta}{((1 - \gamma_s)\eta - n + n_s)} \frac{(\eta - n)}{\eta}$$

$$\times \frac{\pi(n_s, \gamma_s\eta - n_s)\pi(n - n_s, (1 - \gamma_s)\eta - n + n_s)}{\pi(n, \eta - n)}$$

in which the second factor is equal to

$$\frac{B(\gamma_s\eta, (1 - \gamma_s)\eta)}{B(\gamma_s\eta - n_s, (1 - \gamma_s)\eta - n + n_s)}.$$

Substitution of (75) in (16) brings us to

$$(76) \qquad p(_s\ddot{e}_n) = \frac{\gamma_s\eta}{n_s(\gamma_s\eta - n_s)} \frac{(1 - \gamma_s)\eta}{(n - n_s)((1 - \gamma_s)\eta - n + n_s)} \frac{n(\eta - n)}{\eta}$$

$$\times \frac{B(\eta, \eta - n)}{B(n_s, \gamma_s\eta)B(n - n_s, (1 - \gamma_s)\eta - n + n_s)}.$$

5.2. *Generalized Special Values as Weighted Means (continued)*

Substitution of $\eta = -\lambda$ transforms (14) into

$$(76) \qquad p(S/e_n) = (\gamma_s\eta - n_s)/(\eta - n)$$

which may be rewritten as

$$(78) \qquad p(S/e_n) = \frac{\eta}{\eta - n} \gamma_s - \frac{n}{\eta - n} \frac{n_s}{n}.$$

The two factors occurring in (78) cannot be seen as weights: though they sum up to one they are not in the (0, 1)-interval.

It is easy to see that our negative inductive GC-system has the following property

$$(79) \qquad p(S/e_n) < p(S/e'_n) \quad \text{if } n_s > n'_s$$

that is, the larger the number of occurrences of $S$ in $e_n$ the smaller the (generalized) special value. Extending current terminology (see Subsection 4.2) we say that (79) represents the *property* (or principle) *of negative relevance* and for this reason we call a GC-system with this property *negative* inductive.

In order to extend our analysis of GC-systems in Subsection 4.2. in terms of weights, it is convenient to use the originally introduced parameter $\alpha$ and to link $\lambda$ by $\alpha$ according to $\lambda(\alpha) = \alpha/(1 - \alpha)$. As we have seen $\alpha = 0$ corresponds to extreme induction, $0 < \alpha < 1$ to positive induction, $\alpha = 1$ to non-induction and $\alpha > 1$ to negative induction. A GC-system in which $\alpha$ assumes the largest permitted value (viz. $(N - 1)/(N - 1 - \gamma_{\min})$) will be called *maximal negative inductive*.

Consider now the class of GC-systems for which $N$, $w$ and the $\gamma_i$'s are fixed. We say that a member of this class is *stronger* (*weaker*) *inductive than* another member if the $\alpha$-value of the first is smaller (larger) than that of the second. Because this relation orders the class of GC-systems linearly we have that every subset of three different members contains a member, called the *intermediate*, which is stronger inductive than one of the other two but weaker inductive than the third. The following theorem and corollary, for which the proofs require only some algebraic manipulations, bring out the relation between the members of the specified class in a very lucid way.

(T4)　For any set of three GC-systems, with the same values for $N$, $w$ and the $\gamma_i$, the generalized special value in the intermediate system is a weighted mean of the corresponding special values in the two other systems, with weights that have the same structure and that depend only on the $\alpha$'s and $n$, and not on $s$

or $n_s$: more precisely, if $\alpha_1 < \alpha_2 < \alpha_3$, then

$$
\begin{aligned}
(80) \qquad p^{(2)}(S/e_n) &= (n_s + \gamma_s \lambda(\alpha_2))/(n + \lambda(\alpha_2)) \\
&= \frac{\lambda(\alpha_3) - \lambda(\alpha_2)}{\lambda(\alpha_3) - \lambda(\alpha_1)} \cdot \frac{n + \lambda(\alpha_1)}{n + \lambda(\alpha_2)} \cdot p^{(1)}(S/e_n) \\
&+ \frac{\lambda(\alpha_1) - \lambda(\alpha_2)}{\lambda(\alpha_1) - \lambda(\alpha_3)} \cdot \frac{n + \lambda(\alpha_3)}{n + \lambda(\alpha_2)} \cdot p^{(3)}(S/e_n).
\end{aligned}
$$

The factors preceding $p^{(1)}(S/e_n)$ and $p^{(3)}(S/e_n)$ are weights because they are non-negative (as long as $n < N$) and sum up to one; they clearly depend only on the $\alpha$'s and $n$ and they have the same structure.

COROLLARY: the weight of the (generalized special value in the) stronger system increases and the weight of the weaker system decreases by increasing $n$.

Note that the first factors of the weights in (80) do not depend on $n$ and are moreover equal to the weights that result if $\lambda(\alpha_2)$ is written as weighted mean of $\lambda(\alpha_1)$ and $\lambda(\alpha_3)$.

## 5.3 Hypergeometric Systems

Consider a negative inductive GC-system for which there are positive integers $M_i$ such that $M_i = \gamma_i \eta$, for all $i$. It follows that $\eta = \Sigma M_i$ is also a positive integer, say $M$. Now $(10-)$ reduces to

$$(81) \qquad p(Q_i/e_n) = (M_i - n_i)/(M - n)$$

and (72), using (41.2), to

$$(82) \qquad p(e_n) = \left\{ \overline{\cdots} \frac{M_i!}{(M_i - n_i)!} \right\} \Bigg/ \frac{M!}{(M - n)!}.$$

On the basis of (12) we obtain finally

$$(83) \qquad p(\ddot{e}_n) = \left\{ \overline{\cdots} \binom{M_i}{n_i} \right\} \Bigg/ \binom{M}{n}.$$

In the class of urn-model-interpretations (treated in Subsection 3.3) for GC-systems described above, there is one interpretation in which $M$ corresponds directly to the number of balls in the urn and the $M_i$'s to the number of $Q_i$-balls, viz. that one in which $\triangle = -1$ and $\gamma_i = M_i/M$. This is easily verified by comparing (81) with (19). This particular interpretation is usually called random sampling without replacement and the corresponding system of probabilities as the (multiple) hypergeometric distribution.

The parameter-condition $(-)$ tells us now that the $M_i$ may not be smaller than $N - 1$. But let us set aside this condition for a moment. According to (81) $p(Q_i/e_n) = 0$ if $n_i = M_i$ and therefore $p(e_n) = 0$ as soon as there is an $n_i$ larger than $M_i$ and as long as $n$ does not exceed $M$. These observations correspond in the urn-model-interpretation to the fact that it is of course impossible to draw more than $M_iQ_i$-balls or more than $M$ balls in total. The following definition rests on·these observations.

DEFINITION: a hypergeometric system (Hg-system) is a consistent probability pattern w.r.t. $W$. $W^2$. $W^3$. . . .., $W^N$ of a finite set $W$, for which there are integers $M_i$, $0 < M_i < \infty$, $M =_{df} \Sigma M_i \leq N$, such that

(84.1) $\quad p(e_n) = \left\{ \dfrac{M_i!}{\cdots - (M_i - n_i)!} \right\} \bigg/ \dfrac{M!}{(M-n)!}, \quad$ if $n_i \leq M_i$ for all $i$

(84.2) $\quad p(e_n) = 0 (= p(\ddot{e}_n))$. if $n_i > M_i$ for some $i$.

Note that (5) does not define $p(Q_i/e_n)$ in case $p(e_n) = 0$. Therefore the (generalized) special values are only specified by (81) as long as $p(e_n) \neq 0$.

For a particular type of hypergeometric systems it is possible to extend the weight-analysis of the preceding subsection in an interesting way. Let $p_{NM}$ be a particular Hg-system for which there are positive integers $N_i$ such that $NM_i = MN_i$. Hence $\Sigma N_i = N$. Let $M_s$ and $N_s$ be equal to the sum of the $M_i$ and $N_i$ corresponding to the elements of $S$. Let $p_N$ be the Hg-system as $p_{NM}$ with $M_i$ replaced by $N_i$. Of course, $p_{NM}(e_n) \neq 0$ as long as $p_N(e_n) \neq 0$. Suppose that $p_N(e_n) \neq 0$, then we have that, putting $p_0$ for the straight GC-system $(\lambda = 0)$ with $\gamma_i = M_i/M$,

(85) $\quad p_{NM}(S/e_n) = \dfrac{M - N}{N} \dfrac{n}{M - n} \dfrac{n_s}{n} + \dfrac{M N - n}{N} \dfrac{N_s - n_s}{M - n} \dfrac{N_s - n_s}{N - n}$

$\qquad\qquad = \dfrac{M - N}{N} \dfrac{n}{M - n} p_0(S/e_n) + \dfrac{M N - n}{N} \dfrac{1}{M - n} p_N(S/e_n).$

Hence $p_{NM}(S/e_n)$ is the weighted mean of the corresponding generalized special values of $p_0$ and $p_N$. Moreover the weight of $p_0(S/e_n)$ is zero for $n = 0$ and increases to one for $n = N$; conversely, the weight of $p_N(S/e_n)$ is one for $n = 0$ and decreases to zero for $n = N$.

## APPENDIX TO SECTION 2 (PROOF OF T2)

The first part is concerned with the proof of 2(10) (i.e., (10) of Section 2); the second part with the proof of the parameter-condition 2(9). The theorem is trivially true in case $w = 1$; therefore we assume $w = 2, 3, \ldots$.

Let $i, j$ and $k$ range over the indices $1, 2, \ldots, (w)$. Let $n = 0, 1, 2, \ldots, (N - 1)$ and let $n_i = 0, 1, 2, \ldots, n$ such that $\Sigma n_i = n$. We define $f_i(0, 0) = _{\mathrm{df}} \gamma_i$.

According to PIP we may replace 2(7) by

(1.1)    $\gamma_i > 0$         (1.2)    $\sum \gamma_i = 1$

and according to PRR we may replace 2(8) by

(2.1)    $f_i(n, n_i) \geq 0$         (2.2)    $\sum f_j(n, n_j) = 1$.

From (2.2) it follows that

(3)    $f_i(n, n) + \sum_{j \neq i} f_j(n, 0) = 1$    and that

(4)    $f_i(n + 1, 1) + f_k(n + 1, n) + \sum_{j \neq i, k} f_j(n + 1, 0) = 1, \quad i \neq k$.

From POI and PRR we obtain

(5)    $f_i(n, n_i) \cdot f_j(n + 1, n_j) = f_j(n, n_j) \cdot f_i(n + 1, n_i), \quad i \neq j$.

Substitution of $n = n_i = n_j = 0$ in (5) tells us, in combination with (1.1) and (2.1), that $f_i(1, 0)/\gamma_i$ is a non-negative constant (i.e., independent of $i$), say $\alpha$. Therefore we have, using (3) for $n = 1$,

(6.1)    $f_i(1, 0) = \gamma_i \alpha$         (6.2)    $f_i(1, 1) = 1 - \alpha(1 - \gamma_i)$.

Replacement of $\alpha$ by $\lambda/(1 + \lambda)$ in (6) leads to

(7.1)    $f_i(1, 0) = \dfrac{\gamma_i \lambda}{1 + \lambda}$         (7.2)    $f_i(1, 1) = \dfrac{1 + \gamma_i \lambda}{1 + \lambda}$

which is in accordance with 2(10) for $n = 1$.

For the inductive step of the proof we assume first that $w \geq 3$ and that $\alpha \neq 0$ (and therefore $\lambda \neq 0$). Our inductive hypothesis is:

(8)    $f_i(n, n_i) = (n_i + \gamma_i \lambda)/(n + \lambda)$,    for some fixed value of $n$.

From the assumption that $\lambda \neq 0$ it follows that the special values of our inductive hypothesis are all positive and this fact will be used several times implicitly. We start with three applications of (5) in which the assumption that $w \geq 3$ is essential:

(9) $\quad f_k(n + 1, n) = f_i(n + 1, 0) \cdot f_k(n, n)/f_i(n, 0), \quad k \neq i$

(10) $\quad f_j(n + 1, 0) = f_i(n + 1, 0) \cdot f_j(n, 0)/f_i(n, 0), \quad j \neq i$

(11) $\quad f_i(n + 1, 1) = f_j(n + 1, 0) \cdot f_i(n, 1)/f_j(n, 0), \quad i \neq j.$

Substitution of (10) in (11) leads to

(12) $\quad f_i(n + 1, 1) = f_i(n + 1, 0) \cdot f_i(n, 1)/f_i(n, 0).$

If we substitute now (12), (9) and (10) in (4) we obtain

(13) $\quad f_i(n + 1, 0) = f_i(n, 0)/(f_i(n, 1) + f_k(n, n) + \sum_{j \neq i, k} f_j(n, 0))$

and if we substitute here the relevant values of (8) we get

(14) $\quad f_i(n + 1, 0) = \gamma_i \lambda/(n + 1 + \lambda).$

Replacement of $n$ by $n + 1$ in (3) gives

(15) $\quad f_i(n + 1, n + 1) + \sum_{j \neq i} f_j(n + 1, 0) = 1.$

By substitution of the relevant values of (14) in (15) we get

(16) $\quad f_i(n + 1, n + 1) = (n + 1 + \gamma_i \lambda)/(n + 1 + \lambda).$

In order to obtain the remaining special values, i.e., $f_i(n + 1, n_i)$ for $1 \leq n_i \leq n$, we apply (5) once again:

(17) $\quad f_i(n + 1, n_i) = f_j(n + 1, 0) \cdot f_i(n, n_i)/f_j(n, 0), \quad i \neq j$

Substitution of (14), after replacing $i$ by $j$, and (8) in (17) leads to

(18) $\quad f_i(n + 1, n_i) = (n_i + \gamma_i \lambda)/(n + 1 + \lambda), \quad n_i = 1, 2, \ldots, n.$

From (14), (16) and (18) it follows that the inductive step is completed and therefore, because of (7), the proof of 2(10) for $\lambda \neq 0$ and $w \geq 3$.

We turn now to the case $\lambda = 0$ and $w \geq 3$. Of course we need to prove now:

(19) $\quad f_i(n, n_i) = n_i/n, \quad 0 \ll n_i \leq n, n \geq 1.$

For $n = 1$ (19) follows directly from (7). However, we shall see that our inductive step does only work for $n \geq 2$. Hence, we also have to prove as an initial step the case $n = 2$.

Substitution of $n = n_i = 1$ and $n_j = 0$ in (5) gives $f_j(2, 0) = 0$ for all $j$, using that $f_i(1, 1) = 1$ and $f_j(1, 0) = 0$ for all $i$ and $j$. Substitution of this result in (3) leads to $f_i(2, 2) = 1$ for all $i$. Substitution of the same result in (4) leads to $\binom{w}{2}$ equations in $w$ unknowns of the form

$$f_i(2, 1) + f_k(2, 1) = 1, \quad i \neq k.$$

Because $w \geq 3$, the obvious solution of these equations, viz. $f_i(2, 1) = \frac{1}{2}$, is easily seen to be the unique solution: any combination of three of these equations in three unknowns already leads to this particular solution as far as the chosen unknowns are concerned. This completes the proof of (19) for the case $n = 2$.

Consider now (19) for fixed $n \geq 2$ as inductive hypothesis. From substitution of (19) in (5) for $n_i = 0 \neq n_j$ we obtain

(20)    $f_i(n + 1, 0) = 0.$

After replacing $n$ by $n + 1$ in (3), substitution of (20) gives

(21)    $f_i(n + 1, n + 1) = 1$

As an application of (5) we have

(22)    $f_i(n + 1, n_i) = f_j(n + 1, 1) \cdot f_i(n, n_i)/f_j(n, 1), \quad i \neq j$

for $0 \leq n_i < n - 1$, in which the assumption $w \geq 3$ is again essential as soon as $n_i < n - 1$.

Substitution of (19) in (22) transforms (22) into

(23)    $f_i(n + 1, n_i) = n_i \cdot f_j(n + 1, 1), \quad i \neq j, n_i = 0, 1, \ldots, n - 1$

and from this we obtain by substitution of $n_i = 1$

(24)    $f_i(n + 1, 1) = f_j(n + 1, 1).$

Note that the foregoing substitution is not permitted for $n = 1$, but we have excluded this value by assumption.

From (23) and (24) we may conclude

(25)    $f_i(n + 1, n_i) = n_i \cdot c(n + 1), \quad n_i = 0, 1, \ldots, n - 1.$

From substitution of (25) in (2.2) for $n + 1, n_i = n_j = 1$ and $n_k = n - 1$ we obtain

(26)    $c(n + 1) = 1/(n + 1).$

Finally, from (4) we can now get

(27)    $f_i(n + 1, n) = n/(n + 1)$.

Combining (21), (25), (26) and (27) we now have generally

(28)    $f_i(n + 1, n_i) = n_i/(n + 1), \quad n_i = 0, 1, \ldots, n + 1$

and this completes the proof of 2(10) for $\lambda = 0$ and $w \geq 3$.

For the proof of the case that $w = 2$ ($i = 1, 2$) we need SPL, which says in combination with PRR:

(29)    $f_i(n, n_i) = f_i(n, 0) + n_i \cdot g_i(n)$.

Because of (1.2) we have of course that $\gamma_1 = 1 - \gamma_2$ and from (29) and (7) it follows immediately that

(30)    $g_1(1) = 1/(1 + \lambda)$.

Because of (2.2) we have also

(31)    $f_1(n, n_1) + f_2(n, n - n_1) = 1$ and

(32)    $f_1(n + 1, n_1 + 1) + f_2(n + 1, n - n_1) = 1$.

By applying (5) we obtain moreover

(33)    $f_1(n, n_1) \cdot f_2(n + 1, n - n_1) = f_2(n, n - n_1) \cdot f_1(n + 1, n_1)$.

Substitution of (31) and (32) in (33) leads to

(34)    $f_1(n, n_1) \cdot (1 - f_1(n + 1, n_1 + 1))$
        $= (1 - f_1(n, n_1)) \cdot f_1(n + 1, n_1)$.

If we substitute first $n_1 = 0$ in (34) and insert then the relevant applications of (29) we obtain

(35)    $f_1(n, 0) \cdot (1 - g_1(n + 1)) = f_1(n + 1, 0)$.

On the other hand if we substitute first $n_1 = n$ in (34) and insert then the relevant applications of (29) we obtain, after using (35),

(36)    $g_1(n + 1) \cdot (g_1(n) + 1) = g_1(n)$.

It is easy to check that $g_1(n)$ cannot be $-1$, for (36) would then lead to a contradiction. Hence we may rewrite (36) as

(37)    $g_1(n + 1) = g_1(n)/(g_1(n) + 1)$.

Induction on (37), with (30) as initial step, now gives

(38)        $g(n) = 1/(n + \lambda)$

and induction on (35) after substitution of (38), using (7.1),

(39)        $f_1(n, 0) = \gamma_1 \lambda/(n + \lambda)$.

Now it is easy to see that we obtain 2(10) for $i = 1$ from substitution of
(39) and (38) in (29). Finally we get 2(10) for $i = 2$ by using (31), which
completes the proof for the case $w = 2$.

   In the second part of this Appendix we shall prove the parameter-
condition 2(9). Because of (2) the special values are not only required to
sum to unity (2.2) but they also have to be non-negative (2.1). It is easy
to see from 2(10) and the relation between $\alpha$ and $\gamma$ that this condition is
satisfied as long as $0 \leq \alpha \leq 1$. The remaining problem is therefore the
restrictions on $\alpha$ as far as values larger than 1 are concerned, or, in terms
of $\lambda$, restrictions on $\lambda$ as far as (finite) negative values are concerned. In
other words we have to prove according to 2(−):

$$\text{if } 0 < - \lambda = {}_{df}\eta \text{ then } (N - 1)/\gamma_{min} = {}_{df}\eta_{min} \leq \eta.$$

The proof will be given in three steps based on the assumption that
$0 < \eta < \infty$:

(40.1)    $N - 1 < \eta$      (40.2)    $w < \infty$      (40.3)    $\eta_{min} \leq \eta$.

   From 2(10−) and (2.1) follows

(41)        $0 \leq f_i(n, n_i) = (\gamma_i\eta - n_i)/(\eta - n),$   $n_i \leq n = 1, \ldots, (N - 1)$.

If $N$ is infinite then $f_i(n, 0) < 0$ for all $n > \eta$, which is in conflict with (41).
Therefore $N$ has to be finite. If $\eta = N - 1$ then $f_i(N - 1, N_i)$ is undefined,
according to (41). If $\eta < N - 1$ then $f_i(N - 1, 0)$ is negative and therefore
the remaining possibility is (40.1).

   Suppose now that $w = \infty$. It is easy to see that $f_i(N - 1, N - 1)$ is
negative as soon as $\gamma_i < (N - 1)/\eta$ and this is true for arbitrarily many $i$
because the initial special values have to sum to unity. Hence $w$ has to be
finite, and therefore we have also that $\gamma_{min} \neq 0$ and that $\eta_{min} = (N - 1)/$
$\gamma_{min}$ is finite.

   Now it is easy to check that the non-negative condition in (41) holds for
all $i$ and $n_i = 0, 1, \ldots, N - 1$ if and only if $\eta \geq (N - 1)/\gamma_{min} = \eta_{min}$,
i.e., if (40.3) holds.

# CHAPTER 6

# HINTIKKA AND UNIVERSALIZED CARNAPIAN SYSTEMS

## 1. INTRODUCTION

In Section 3.6 of the preceding chapter we have seen that the class of proper Carnapian systems is that subclass of GC-systems whose members satisfy the principle of initial equipossibility. It is easy to see that the conjunction of this principle and the principle of restricted relevance is equivalent to:

(PERR) *principle of equal restricted relevance*
$$p(Q_i/e_n) = f(n, n_i).$$

As a special case of T3 (Section 4.2) of the preceding chapter we have that, in a positive inductive C-system, if $S$ is a non-empty subset of $W$ then $p(SSS\ldots/e_n) = 0$. In the light of Carnap's own interpretation this amounts to zero (prior and posterior) probability for any universal statement claiming that all individuals of a (denumerably) infinite universe have the property designated by S. Carnap came to the conclusion ([2], p. 575) that universal statements (including universal laws) had to be considered as non-essential for pure (and applied) science. According to him we only need 'instance-confirmation-values': with respect to the described universal hypothesis this is the value $p(S/e_n)$.

In this chapter we shall see how it is possible to escape these highly problematic conclusions.

## 2. NH-SYSTEMS

In this section we shall study the class of systems introduced by Hintikka and Niiniluoto [19]. These systems were anticipated by me in a rather clumsy way [23]. The content of this section is essentially contained in the paper of Hintikka and Niiniluoto, but our presentation will be rather different.

Let $K$ be a finite set of $k > 2$ elements, called 'Q-predicates', and let $e_n$ be an element and $E_n$ a subset of $K^n$. With respect to subsets $W$ of $K$ we assume the convention $|W| = w$. As in the preceding chapter we are interested in consistent probability patterns, but now always with respect to the infinite sequence $K, K^2, K^3, \ldots$ and we assume in addition that these patterns are regular, i.e., $p(e_n)$ is always positive. (Note that the latter assumption implies PIP.) In sum we have therefore for all (finite) $n$:

(1.1)     $p(e_n) > 0$         (1.2)     $\displaystyle\sum_{e_n \in K^n} p(e_n) = 1$

(2)     $p(E_n) = \displaystyle\sum_{e_n \in E_n} p(e_n)$

(3)     $p(E_n) = p(E_n K)$.

The theorem of Kolmogorov guarantees again that the definition

(4)     $p(E_n KKKK \ldots) =_{\mathrm{df}} p(E_n)$

leads to a unique probability measure on the set of (measurable) subsets of $KKK. \ldots$ (All subsets to be considered are measurable.)

From the regularity assumption it follows that the definition

(5)     $p(Q_i/e_n) =_{\mathrm{df}} p(e_n Q_i)/p(e_n)$

is always appropriate. In what follows, expressions of the form '$p(\ldots/e_0)$' are intended to be equivalent to '$p(\ldots)$'. The product rule, i.e., repeated application of

(6)     $p(e_n Q_i) = p(e_n)p(Q_i/e_n)$

shows that a pattern is completely determined as soon as we have all its special values $p(Q_i/e_n)$. These values are bound by the restrictions

(7.1)     $p(Q_i/e_n) > 0$         (7.2)     $\displaystyle\sum_{Q_i \in K} p(Q_i/e_n) = 1.$

With respect to a fixed $e_n$ we again use the convention that $n_i$ indicates the number of occurrences of $Q_i$ in $e_n$. In addition $M$ indicates the set of $Q_i$'s for which $n_i > 0$; and $|M| = m$. If we want to make explicit that $n_i$ and $M$ are related to $e_n$ we write $n_i(e_n)$ and $M(e_n)$ respectively. By $e_m, m = 1, 2, \ldots, k$, we indicate of course a member of $K^m$ for which $|M(e_m)| = m$.

Consider now a regular consistent probability pattern satisfying

(POI)     *principle of order indifference*
          $p(Q_i Q_j/e_n) = p(Q_j Q_i/e_n)$

and

(WPERR) *weak principle of equal restricted relevance*
$$p(Q_i/e_n) = f_m(n, n_i).$$

The latter principle differs from the Carnapian PERR (see the Introduction) in that the size of $M(e_n)$ may play a role in addition to $n$ and $n_i$.

The following conventions will appear to be useful. We may replace $p(\overline{M}/e_n)$, which is equal to $(k - m)f_m(n, 0)$, by $h(n, m)$ and $p(M/e_n)$ $(= 1 - h(n, m))$ by $g(n, m)$. Of course we have $h(0, 0) = 1$ and $h(n, k) = 0$ for all $n \geq k$; moreover $h(n, m)$ and $g(n, m)$ are only defined for $n = m = 0$ and $m = 1, 2, \ldots, \min(n, k)$.

From (7) we may conclude

(8)      $0 < h(n, m) < 1$      $0 < m \leq \min(n, k - 1)$.

On the basis of (8) it is easy to show now that WPERR is equivalent to the conjunction of the following principles:

(NH1)    $p(M/e_n) = g(n, m)(=_{df} 1 - h(n, m))$

(NH2.1) $p(Q_i/e_n\overline{M}) = (p(Q_i/e_n)/p(\overline{M}/e_n)) = 1/(k - m)$,    $Q_i \notin M$

(NH2.2) $p(Q_i/e_nM) = (p(Q_i/e_n)/p(M/e_n)) = k_m(n, n_i)$,    $Q_i \in M$.

The following theorem can now be proved:

(T1)      If a regular consistent probability pattern satisfies POI and the NH-principles, then there is a real number $\rho$, $-1 < \rho \leq \infty$, such that

(9)      $p(Q_i/e_nM) = k_m(n, n_i) = \dfrac{n_i + \rho}{n + m\rho}$,    $Q_i \in M$.

The proof will be given in the Appendix to this chapter.

We introduce now the following:

DEFINITION: an NH-system is a regular consistent probability pattern w.r.t. $K, K^2, K^3, \ldots (2 \leq k < \infty)$ satisfying POI, NH1 and NH2.1 for which there is a real number $\rho$, $0 < \rho < \infty$ such that (9) holds.

In this definition we have also included the case $k = 2$; for a complete axiomatic derivation of this particular case we need a linearity assumption for $k_2(n, n_i)$, $0 < n_i < n$, similar to SPL. Notice that we have excluded the cases $\rho \leq 0$. This comes about to the requirement that either $p(Q_i/e_nM)$ is uniformly equal to $1/m$ (the case $\rho = \infty$) or $n_i/n \gtreqless p(Q_i/e_nM) \gtreqless 1/m$ iff $n_i/n \gtreqless 1/m$ ($0 < \rho < \infty$). It is not altogether clear whether systems with $-1 < \rho < 0$ are probabilistic; how this may be is not of interest for our purposes. In a different context we shall pay some attention to systems with $\rho = 0$ (see Section 7.3).

The following theorem will appear to have important consequences:

(T2)      In an NH-system holds, for $m = 1, 2, \ldots, \min (n, k) - 1$,

$$(10) \qquad h(n + 1, m) = \frac{h(n, m)}{1 - h(n, m)} \cdot (1 - h(n + 1, m + 1)) .$$

$$\cdot \frac{n + m\rho}{n + 1 + (m + 1)\rho} .$$

*Proof:* From POI, NH1, NH2.1 and (9) follows, for fixed $e_n$, $Q_i \in M$, $Q_j \notin M$,

$$p(Q_iQ_j/e_n) = (1 - h(n, m)) \cdot \frac{n_i + \rho}{n + m\rho} \cdot h(n + 1, m) \cdot \frac{1}{k - m}$$

$$= p(Q_jQ_i/e_n) = h(n, m) \cdot \frac{1}{k - m} \cdot (1 - h(n + 1, m + 1)) .$$

$$\cdot \frac{n_i + \rho}{n + 1 + (m + 1)\rho} ., \; q.e.d.$$

Suppose now that the values of $\rho$, $h(1, 1)$, $h(2, 2)$, $\ldots$, $h(k - 1, k - 1)$ are given. From (10) it follows directly that all $h(n, k - 1), n > k - 1$, can be calculated successively, starting with $h(k, k - 1)$ (use: $h(n + 1, k)$ is zero). We need for these calculations only $\rho$, $h(k - 1, k - 1)$. Suppose now that we have calculated, for fixed $m + 1 < k$, all $h(n + 1, m + 1), n > m$, then we obtain from (10) all $h(n, m), n > m$. For this calculation we need $h(m, m)$, in addition to $\rho$, $h(k - 1, k - 1)$, $\ldots$, $h(m + 1, m + 1)$.

We may conclude therefore that our fixed values determine the systems completely. This holds also if we fix the values $\rho$, $g(1, 1)$, $g(2, 1)$, $\ldots$, $g(k - 1, 1)$, as is seen by re-writing (10) in terms of '$g(. , .)$' as an equation for $g(n + 1, m + 1)$. Then it is easily seen to be possible to calculate successively $g(2, 2)$, $\ldots$, $g(k - 1, 2)$, $g(3, 3)$, $\ldots$, $g(k - 1, 3)$, $\ldots g(k - 2, k - 2)$, $g(k - 1, k - 2)$, $g(k - 1, k - 1)$. In particular we have now $g(m, m)$ for $m = 1, 2, \ldots, k - 1$ which brings us back to the previous case. Hence we have

(T3)    An NH-system can be determined completely by
$k$ parameters, e.g.,
$\rho, h(m, m), \quad m = 1, 2, \ldots, k - 1, \quad$ or
$\rho, g(m, 1), m = 1, 2, \ldots, k - 1.$

It is important to note that T3 does not imply that any choice of the parameters in accordance with (8) is adequate (see Section 7.1). We call the parameters mentioned in T3 *finite* because they are defined on finite Cartesian products of $K$.

## 3. HINTIKKA-SYSTEMS (H-SYSTEMS)

Hintikka [16] has introduced a regular consistent probability pattern w.r.t. $K$, $K^2$, $K^3$, ... based on 'relative' patterns. Here we shall give its axiomatic introduction. We need the following definitions: let $W \subset K$, $W \neq \varnothing$,

(11)    $H_W(n) =_{\mathrm{df}} \{e_n \in K^n / M(e_n) = W\}.$

Notice that $H_W(n)$ is a subset of $W^n$ and, moreover, that it is non-empty if and only if $w \leq n$.

(12)    $H_W =_{\mathrm{df}} \bigcup_{n=w}^{\infty} H_W(n) W W' W'. \ldots$

$H_W$ is a (measurable) subset of $W^\infty$. The $H_W$ constitute a partition of $K^\infty$:

(13.1)    if $W \neq V$ then $H_W \cap H_V = \varnothing$    (13.2)   $\bigcup_{W \subset K} H_W = K^\infty.$

Let there be a probability function $q$ on the set of $H_W$ ($\varnothing \neq W \subset K$):

(14.1)    $0 \leq q(H_W)$    (14.2)   $\sum_{W \subset K} q(H_W) = 1$

such that, moreover,

(14.3)    $q(H_K) > 0.$

Let there also be, for each $H_W$, a so-called relative or conditional regular consistent probability pattern w.r.t. $W$, $W^2$, $W^3$, ...: $q_W$, i.e., for $e_n \in W^n$, $E_n \subset W^n$

(15.1)    $q_W(e_n) > 0$    (15.2)   $\sum_{e_n \in W^n} q_W(e_n) = 1$

(16)     $q_W(E_n) = \sum\limits_{e_n \in E_n} q_W(e_n)$

(17)     $q_W(E_n) = q_W(E_n W)$.

On the basis of $q$ and the $q_W$'s we define for each $E_n \subset K^n$:

(18)     $p(E_n) = \sum\limits_{e_n \in E_n} \sum\limits_{W \supset M(e_n)} q(H_W)q_W(e_n)$.

It is easy to see that $p$ is a regular consistent probability pattern w.r.t. $K$, $K^2$, $K^3$, . . . .

Application of Bayes' rule gives us as a posterior probability function on the $H_W$:

(19)     $q(H_W/e_n) = q(H_W)q_W(e_n)/p(e_n)$.

We shall use the abbreviation '$H_W e_n$' for the set $H_W \cap e_n KKK \ldots$, which is equal to $H_W \cap e_n WWW. \ldots$ Kolmogorov's extension theorem guarantees that, for $V \subset W$ and $e_n \in W^n$, $q_W(H_V)$ and $q_W(H_V e_n)$ are uniquely determined and therefore also $q_W(H_V/e_n)$ and $q_W(e_n/H_V)$, provided that $q_W(H_V) \neq 0$.

With regard to $p$ Kolmogorov's theorem guarantees that, for all $W$ and $e_n \in K^n$, $p(H_W)$ and $p(H_W e_n)$ are uniquely determined and therefore also $p(H_W/e_n)$ and $p(e_n/H_W)$, provided that $p(H_W) \neq 0$.

We lay down first restrictions to the relative patterns. In fact we shall assume that they are regular Carnapian systems depending only on the size of $W$: i.e., let $e_n \in W^n$; $Q_i$, $Q_j \in W$:

(H1)     $q_W(Q_i/e_n) = f^w(n, n_i)$                (equal restricted relevance)

(H2)     $q_W(Q_i Q_j/e_n) = q_W(Q_j Q_i/e_n)$   (order indifference)

For the special case $|W| = 2$ we require in addition the special principle of linearity (see Chapter 5, Section 2).

From T2 (for infinite $N$) of Chapter 5, Section 2, and the regularity assumption, we may conclude that there are, for all $w = 2, 3, \ldots, k$, real numbers $\rho_w$, $0 < \rho_w \cdot \infty$, such that

(20)     $q_W(Q_i/e_n) = (n_i + \rho_w)/(n + w\rho_w)$.

On the basis of T3 of Chapter 5, Section 4.2, we may now infer the following important consequence:

(21.1)   $q_W(H_W) = 1$              (21.2)  $q_W(H_V) = 0$ if $V \subset W$, $V \neq W$.

In the theorem referred to, we proved in fact that $q_W(VVV\ldots) = 0$ if $V \subset W$, $V \neq W$. $H_V$ is a subset of $V^\infty$, therefore (21.2). From $1 = p_W(W^\infty) = \sum_{V \subset W} p_W(H_V)$ we obtain now (21.1).

Important consequences of (21) are:

(22)    $p(H_W) = q(H_W)$

(23)    $p(e_n/H_W) = q_W(e_n)$,   $p(H_W) \neq 0$

and as a generalization of (22)

(24)    $p(H_W/e_n) = q(H_W/e_n)$.

*Proof:*    From (18) we may infer

$$p(H_W) = \sum_{V \supseteq W} q(H_V)q_V(H_W) \overset{(21)}{==} q(H_V), \text{ hence (22),}$$

and also

$$p(H_W e_n) = \sum_{V \supseteq W} q(H_V)q_V(H_W e_n).$$

Now, if $V \supset W$ and $V \neq W$, then $0 \leq q_V(H_W e_n) \leq q_V(H_W) \overset{(21)}{==} 0$, therefore, using (22),

$$p(H_W e_n) = p(H_W)q_W(H_W e_n).$$

By definition we have also, if $p(H_W) \neq 0$,

$$p(H_W e_n) = p(H_W)p(e_n/H_W).$$

Hence $p(e_n/H_W) = q_W(H_W e_n)$.

Consider now

$$q_W(e_n) = q_W(e_n WWWW\ldots)$$
$$= q_W(H_W e_n) + q_W((H^\infty - H_W) \cap e_n WWW\ldots).$$

The second term is (non-negative and) $\leq q_W(W^\infty - H_W) = 1 - q_W(H_W) \overset{(21)}{==} 0$, hence $q_W(e_n) = q_W(H_W e_n)$ and the latter was equal to $p(e_n/H_W)$ which completes the proof of (23). Now (24) follows directly from (19), (22) and (23), Q.E.D.

From these consequences it follows of course that we may avoid from now on the use of $q$-expressions, and even that it would have been adequate if we had started directly with $p$-expressions. Moreover, according to H1, we may use '$H_w$' in the context of a conditional pattern to indicate an arbitrary $H_W$ (with $|W| = w$). In the following results, which are easy to prove and essentially contained in the preceding chapter, it is always presupposed that $e_n$ and $Q_i$ are compatible with $H_w$.

$$(25.1) \quad p(M/H_w e_n) = \frac{n + m\rho_w}{n + w\rho_w}$$

$$(25.2) \quad p(\overline{M}/H_w e_n) = \frac{(w - m)\rho_w}{n + w\rho_w}$$

$$(26.1) \quad p(Q_i/H_w e_n M) = \frac{n_i + \rho_w}{n + m\rho_w} \quad (Q_i \in M)$$

$$(26.2) \quad p(Q_i/H_w e_n \overline{M}) = \frac{1}{(w - m)} \quad (Q_i \notin M)$$

$$(27) \quad p(e_n/H_w) = \frac{\prod_i \pi(n_i, \rho_w)}{\pi(n, w\rho_w)}$$

(in which $\pi(k, x) = x(x + 1) \ldots (x + k - 1)$ if $k = 1, 2, \ldots$ and $\pi(0, x) = 1$).

Our next step is a strong restriction on the probability function on the $H_W$:

(H3)     $p(H_W) = p(H_V)$   if $v = w$.

In other words, the $H_W$ get the same value if their constitutive sets have equal size. By consequence we may write $p(H_W) = p(H_w)$. Now we are in a position to introduce the following

DEFINITION: an H-system is a regular consistent probability pattern w.r.t. $K, K^2, K^3, \ldots$, $(2 \leq k < \infty)$ generated in the described way ((14)–(19)), satisfying H3, and for which there are, for each $w = 2$, $3, \ldots, k$, real numbers $\rho_w$, $0 < \rho_w \leq \infty$, such that (20) holds. A Special H-system, an SH-system, is an H-system in which all $\rho_w$ are equal to, say, $\rho$.

It is easily seen from (18) and (20) that the following theorem holds:

(T4)  The values $p(H_w)$, $w = 1, 2, \ldots, k - 1$ and $\rho_w$, $w = 2, 3, \ldots, k$ constitute a complete set of $(2k - 2)$ parameters for an H-system. It is necessary and sufficient that they are real values such that $p(H_w) \geq 0$, $\sum_{w=1}^{k-1} \binom{k}{w} p(H_w) < 1$ and $0 < \rho_w \leq \infty$. If $p(H_w) = 0$ for $w = 1, 2, \ldots, k - 1$, and hence $p(H_k) = 1$, the resulting H-system is in fact a C-system with $\lambda = \lambda_k = k\rho_k$.

## 4. SOME FUNDAMENTAL PROPERTIES OF H-SYSTEMS

H-systems satisfy a number of requirements which one might consider to be desirable in certain contexts. We start, however, with a property which holds under very general conditions.

(T5)  universal confirmation

(28)  $p(H_M/e_n M) > p(H_M/e_n)$,  $p(H_M) > 0$, $M \neq K$

*Proof:*  We may rewrite (28) as

$$p(H_M e_n M)/p(e_n M) > p(H_M e_n)/p(e_n).$$

Now, because $p(H_M e_n M) = p(H_M e_n)$ (for $H_M$ implies $M$), this inequality holds as soon as $p(H_M e_n) > 0$ and $p(M/e_n) < 1$. In an H-system both conditions are satisfied if $p(H_M) > 0$ and $M \neq K$,   *q.e.d.*

In Section 11 we shall introduce the class of so-called GH-systems, which contains the H-systems as special cases. The following property turns out to hold for all GH-systems. For that reason we may formulate the proof in general terms. This gives us, in addition, the opportunity to refer to this proof as a hint for the way in which proofs of so-called (strong) universal-instance confirmation (see below) have to be set up.

(T6)  instantial confirmation (positive instantial relevance)

(29)  $p(Q_i/e_n Q_i) > p(Q_i/e_n)$
      if $M = K$, only if $\rho_K < \infty$, and, if $M \neq K$ and $\rho_K = \infty$, only if $p(H_W) > 0$ for some $W \supset M$, $W \neq K$.

*Proof:*  In this proof we write $p_W(e_n)$ instead of $p(Hwe_n)$. Read as a proof for H-systems, $\rho_W$ may be replaced by $\rho_w$. We assume first that

$p(H_W) > 0$ for all $W \supset M$. We may rewrite (29) as

(i) $\qquad p(e_n)p(e_nQ_iQ_i) > \{p(e_nQ_i)\}^2.$

Suppose first that $Q_i \in M$. By (18), (22) and (23) we may rewrite (i) as

(ii) $\qquad \sum\limits_{W \supset M} \sum\limits_{V \supset M} p(H_W)p(H_V)p_W(e_n)p_V(e_n)p_V(Q_iQ_i/e_n)$

$\qquad > \sum\limits_{W \supset M} \sum\limits_{V \supset M} p(H_W)p(H_V)p_W(e_n)p_V(e_n)p_W(Q_i/e_n)p_V(Q_i/e_n).$

For the terms with $V = W$ we have by (20), if $\rho_W < \infty$

(iii) $\qquad p_W(Q_iQ_i/e_n) > \{p_W(Q_i/e_n)\}^2.$

For any pair $(W, V)$ of terms with $V \neq W$ and $\rho_W < \infty$ we have with (iii)

(iv) $\qquad p_W(Q_iQ_i/e_n) + p_V(Q_iQ_i/e_n) > \{p_W(Q_i/e_n)\}^2 + \{p_V(Q_i/e_n)\}^2$
$\qquad \geq 2p_W(Q_i/e_n)p_V(Q_i/e_n).$

This gives (29) if $\rho_W < \infty$ for some $W \supset M$. If $\rho_W = \infty$ for all $W \supset M$, then the first member of (iv) is equal to the second, but if $V \neq W$, the second is larger than the third. Hence, (29) turns into an equality only if $M = K$ and $\rho_K = \infty$.

If $Q_i \notin M$, and hence $M \neq K$, the summation in (ii) may be taken over $W \supset M$, $Q_i \in W$ and $V \supset M$, $Q_i \in V$ respectively (called the main sums) such that an additional sum appears at the left side, viz.:

$$\sum\limits_{\substack{W \supset M \\ Q_i \notin W}} \sum\limits_{\substack{V \supset M \\ Q_i \in V}} p(H_W)p(H_V)p_W(e_n)p_V(e_n)p_V(Q_iQ_i/e_n).$$

It is easy to see that the main sum at the left side is then not smaller than the (main) sum at the right. Hence (29) holds as soon as the additional sum is positive, which is the case (take, e.g., $W = M$ and $V = K$). This completes the proof under the assumption that $p(H_W) > 0$ for all $W \supset M$.

Note that $p(H_K) > 0$ by definition; hence, there remains to be studied the case that $M \neq K$ and $p(H_W) = 0$ for some $W \supset M$, $W \neq K$. Consider again first $Q_i \in M$. By (iv) we may conclude now to (29) as soon as $\rho_W < \infty$ and $p(H_W) > 0$ for some $W \supset M$, which is already the case if $\rho_K < \infty$. If $\rho_K = \infty$ and $p(H_W) = 0$ for all $W \supset M$, $W \neq K$, then (ii), and therefore (29), turns into an equality. If $\rho_K = \infty$ and $p(H_W) > 0$, for some $W \supset M$, $W \neq K$, then the first member of (iv) is larger than the third for the pair $(W, K)$, and this leads to (29).

Finally, if $Q_i \notin M \neq K$ and $p(H_W) = 0$ for some $W \supset M$, $W \neq K$, the above indicated main sums guarantee (29) as soon as $\rho_K < \infty$. If $\rho_K = \infty$ and $p(H_W) = 0$ for all $W \supset M$, $W \neq K$, the additional sum is zero and the main sums are equal; hence, in this case, (29) turns into an equality. Now, let $\rho_K = \infty$ and $p(H_W) > 0$ for some $W \supset M$, $W \neq K$. If $Q_i \notin W$, the additional sum is positive, which leads to (29). If $Q_i \in W$, then the first member of (iv) is larger than the third for the pair $(W, K)$. Hence, the main sum at the left is larger than the main sum at the right side, $Q.E.D.$

The proof of the following property is similar to that of the previous one (for $Q_i \in M$).

(T7)    universal-instance confirmation

(30)    $p(M/e_n M) > p(M/e_n)$,    $M \neq K$
        if $\rho_K = \infty$, only if $p(H_W) > 0$ for some $W \supset M$, $W \neq K$.

We conclude this survey of inequalities of H-systems by stating, without proof, that the following inequalities do *not* hold generally:

(31)*    $p(H_M/e_n Q_i) > p(H_M/e_n)$,    $Q_i \in M, p(H_M) > 0, M \neq K$
         (called: strong universal confirmation)

(32)*    $p(M/e_n Q_i) > p(M/e_n)$,    $Q_i \in M, M \neq K$
         (called: strong universal-instance confirmation).

Suppose that we want to apply an H-system to an infinite sequence of experiments with possible outcomes in $K$. Suppose further that it turns out in the long run that only members of a subset $M$ of $K$ occur as outcomes. In this situation, i.e., if $n \to \infty$ and $m$ remains constant from a certain point on, we might want to have that $p(H_M/e_n)$ goes to 1 in a sense which has still to be made precise. The next theorem states that an H-system has this property in a rigorously defined way.

(T8)    universal convergence
        $p(H_M/e_n^M) \to 1$ if $n \to \infty$ and $M$ remains constant in the sense
        that, if $M(e_n) = M$, then $\lim_{n \to \infty} p(H_M/e_n M^s) = 1$ and, for
        $m < w \cdot k$, $\lim_{s \to \infty} p(H_w/e_n M^s) = 0$, provided that $p(H_M) > 0$.

*Proof:*  $p(H_M/e_n M^s) = \dfrac{p(H_M)p(e_n/H_M)p(M^s/H_M e_n)}{\displaystyle\sum_{w=m}^{k} \binom{k-m}{w-m} p(H_w)p(e_n/H_w)p(M^s/H_w e_n)}$.

From T3 of Chapter 5, Section 4.2 it follows directly, using (25), that $\lim_{s \to \infty} p(M^s/H_w e_n)$ is 0 if $w > m$ and is 1 if $w = m$. Hence $p(H_M/e_n M^s) \to 1$

if $s \to \infty$. That $p(H_w/e_n M^s) \to 0$ for all $w > m$ follows now from the fact that they are all non-negative and that

$$\sum_{w=m+1}^{k} \binom{k-m}{w-m} p(H_w/e_n M^s) = 1 - p(H_M/e_n M^s), \quad q.e.d.$$

A direct consequence of T8 is

(T9)     universal-instance convergence
         $p(M/e_n) \to 1$ if $n \to \infty$ and $M$ remains constant in the sense
         that $\lim_{s \to \infty} p(M/e_n M^s) = 1$ provided that $p(H_M) > 0$.

We conclude the analysis of the limit behaviour by

(T10)    instantial (or relative frequency) convergence
         $p(Q_i/e_n)$ tends to the relative frequency $n_i/n$ if $n \to \infty$ in the
         sense that $\lim_{n \to \infty} p(Q_i/e_n) - n_i/n = 0$ if all $p$-parameters are finite.

*Proof:*   $p(Q_i/e_n) - n_i/n = \sum_{w=m}^{k} \binom{k-m}{w-m} p(H_w/e_n)(p(Q_i/H_w e_n) - n_i/n)$

because the sum equals 1 if we omit the last factor of each term. Now it follows directly from (20) that

$$\lim_{n \to \infty} (p(Q_i/H_w e_n) - n_i/n) = 0 \text{ if } \rho_w \text{ is finite}, \quad q.e.d.$$

From the proof of this theorem we may conclude that the convergence of $p(Q_i/e_n)$ to $n_i/n$ 'holds everywhere', whereas, e.g., in case of repeated experiments with constant (objective) probabilities $q_i$ for $Q_i$, the convergence of $n_i/n$ to $q_i$ is only 'with objective probability 1'.

It is easy to check that Carnapian systems have, in case $0 < \lambda = k\rho < \infty$, the properties of (strong) universal-instance and instantial confirmation as well as the properties of universal-instance and instantial convergence, but they do not have the properties of (strong) universal confirmation and universal convergence, except that the latter property holds trivially in case $M = K$.

## 5. AN URN-MODEL FOR H-SYSTEMS

It is not difficult to see how the urn-model of Chapter 5, Section 3.3 can be modified in such a way that it becomes an interpretation (a model) for H-systems. Let $K$ be a set of (different) colours and let there be for each non-empty subset $W$ of $K$ an urn $U_W$ containing $N$ balls of colour $Q_i$ for all $Q_i \in W$. We say that $U_W$ instantiates the colours of $W$.

Suppose someone $(X)$ selects one urn with probabilities $P(U_W)$ satisfying (the relevant reformulations of) (14) and H3. Hence we may also write $P(U_w)$ instead of $P(U_W)$.

Suppose that we, as investigators, do not come to know which urn has been drawn, but only the $P(U_w)$ used by $X$. Suppose further that $X$ draws randomly one by one a ball out the selected urn. He tells us the colour of the drawn ball, puts it back and adds a fixed number of new balls with the same colour into the urn. We have only the following conditional information about this number: if $X$ is drawing in an urn instantiating $w$ colours he adds $\Delta_w$ balls.

Now let $X$ have performed $n$ experiments with (ordered) result $e_n$. Of course we may conclude that he can only be drawing in an urn instantiating at least the members of $M(e_n)$. Moreover, if he is in fact drawing in $U_w(w \geq m)$ then the probability that the colour of the next ball will be $Q_i$, $P_w(Q_i/e_n)$, is obviously $(n_i\Delta_w + N)/(n\Delta_w + wN)$, which is equal to $(n_i + \rho_w)/(n + w\rho_w)$ if we replace $N/\Delta_w$ by $\rho_w$.

It will be clear now that this probability process, from our point of view, corresponds with an H-system in which $\rho_w = N/\Delta_w$ and $P(U_w) = p(H_w)$.

Notice that this model is particularly easy to imagine if $X$ selects the urn randomly: there are $2^k - 1$ urns and hence $P(U_w)$ becomes $1/(2^k - 1)$. Another simple possibility is that $X$ selects first, with equal probability, the number of colours $(1, 2, \ldots, k)$ in the urn to be selected and then, again with equal probability, one urn out of the set of urns with the selected number of colours, hence $P(U_w) = (1/k) \cdot 1/\binom{k}{w}$.

In case of an SH-system $\Delta_w$ is the same for all $w$ and the entire process can be described without the introduction of $X$: we may select an urn and perform the drawings (without looking in the urn), for now it is possible to add new balls adequately without knowing what type of urn it is.

Consider finally the case that $\Delta_w = \rho_w = N = 1$. Each urn contains, at the beginning, one ball of each colour it instantiates and one $Q_i$-ball is added if a $Q_i$-ball is selected.

## 6. THE EQUIVALENCE OF NH- AND SH-SYSTEMS: UNIVERSALIZED CARNAPIAN SYSTEMS (UC-SYSTEMS)

In Section 3 we have defined SH-systems as H-systems in which all $\rho_w$ are equal to a fixed number $\rho$. In an SH-system we have therefore, as special cases of (20), (25) and (26) for $w \geq m$:

(33)        $p(Q_i/H_w e_n) = \dfrac{n_i + \rho}{n + w\rho}$

(34.1)      $p(M/H_w e_n) = \dfrac{n + m\rho}{n + w\rho}$

(34.2)      $p(\overline{M}/H_w e_n) = \dfrac{(w - m)\rho}{n + w\rho}$

(35.1)      $p(Q_i/H_w e_n M) = \dfrac{n_i + \rho}{n + m\rho}$   $(Q_i \in M)$

(35.2)      $p(Q_i/H_w e_n \overline{M}) = \dfrac{1}{(w - m)}$   $(Q_i \notin M)$

(T11)   A consistent probability pattern $p$ w.r.t. $K, K^2, K^3, \ldots, k > 2$, is an SH-system if (and only if) $H_1$, $H_2$, $H_3$ and $p(H_k) > 0$ are satisfied, and in addition

(H4)    $p(Q_i/H_w e_n M) = p(Q_i/H_v e_n M)$   $Q_i \in M, w \geq m, v \geq m$

(H4.1)  $f^2(1, 1) = f^2(3, 1)/(4f^2(3, 1) - 1)$   $(f^w(n, n_i) = p(Q_i/H_w e_n))$
        (hence: the special principle of linearity is not required).

*Proof:*  If $w > 2$ we can derive (20) from H1 and H2, and therefore (26). Now H4 and (26) imply together that the $\rho_w$'s $(w > 2)$ are constant. Hence we may conclude to (35.1) for $w > 2$. But then H4 implies (35.1) also for $w = 2$ (and $m = 1$ or 2). If $m = w = 2$, then (35.1) implies (33) for in this case 'H$_2$ implies $M$'. The only thing which still needs to be proved therefore is that (33) holds for $w = 2$ and $m = 1$. In other words we have to prove that $f^2(n, n) = (n + \rho)/(n + 2\rho)$ or, which clearly amounts to the same thing, that $f^2(n, 0) = \rho/(n + 2\rho)$. Now from H1 and H2 it follows that the equation $f^2(n, n)f^2(n + 1, 0) = f^2(n, 0)f^2(n + 1, n)$. Moreover, we already know that $f^2(n + 1, n) = (n + \rho)/(n + 1 + 2\rho)$. Suppose now for a moment that $f^2(n, n) = (n + \rho)/(n + 2\rho)(= 1 - f^2(n, 0))$ for some particular value of $n$. Our equation implies then that $f^2(n + 1, 0) = \rho/(n + 1 + 2\rho)$. Therefore it is sufficient to assume that $f^2(1, 1) = (1 + \rho)/(1 + 2\rho)$ and this is guaranteed by the special axiom H4.1, *Q.E.D.*

It is easy to see that T11 does not hold for the case that $k = 2$ for H4 cannot do the job described in the proof for the case $k > 2$. Therefore, we need to assume the special principle of linearity. Notice also that there is in this case only one $\rho$-parameter.

The main part of this section will consist of the proof of an equivalence theorem to the effect that all SH-systems are NH-systems and vice versa. The NH-systems were introduced in Section 2 on the basis of POI and the weak principle of equal restricted relevance (WPERR). The one side of the theorem is not difficult to prove:

(T12.1)   An SH-system is an NH-system.

*Proof:*   From the fact that the conditional patterns of an H-system satisfy POI, it follows directly that H-systems, and therefore SH-systems, satisfy POI. In order to get more insight into SH-systems we shall prove the fulfilment of WPERR indirectly through the NH-principles. (A direct proof is in fact much shorter.)

(a)      Consider first, for $Q_i \in M$,

$$p(Q_i/e_n M) = \sum_{w=m}^{k} \binom{k-m}{w-m} p(H_w/e_n M)p(Q_i/H_w e_n M).$$

Note that the right-hand sum equals 1 if the last factor in each term is replaced by 1. With (35.1) it follows that,

(36)     $p(Q_i/e_n M) = (n_i + \rho)/(n + m\rho)$   $Q_i \in M$.

This does not only imply that NH2.2 is satisfied but also that $p(Q_i/e_n M)$ gets the same value in an SH-system as in an NH-system as soon as their $\rho$-parameters are identical. Compare (36) with (9).

(b)      Consider next, for $Q_i \notin M$,

$$p(Q_i/e_n \overline{M}) = \sum_{w=m+1}^{k} \binom{k-m-1}{w-m-1} p(H_w/e_n \overline{M})p(Q_i/H_w e_n \overline{M}).$$

Using (35.2) we get now directly NH2.1:

(37)     $p(Q_i/e_n \overline{M}) = 1/(k-m)$   $Q_i \notin M$.

(c)      For the proof of NH1 note first that (27) reduces to

(38)     $p(e_n/H_w) = \left\{ \prod_i \pi(n_i, \rho) \right\} / \pi(n, w\rho)$

and therefore we have:

$$(39) \qquad p(e_n) = \left\{ \prod_i \pi(n_i, \rho) \right\} \sum_{w=m}^{k} \binom{k-m}{w-m} p(H_w)/\pi(n, w\rho).$$

On the basis of (19) we obtain now:

$$(40) \qquad p(H_w/e_n) = \left( p(H_w)/\pi(n, w\rho) \right) \cdot \left( \sum_{v=m}^{k} \binom{k-m}{v-m} p(H_v)/\pi(n, v\rho) \right).$$

(40) leads together with (34.1) to:

$$(41) \qquad p(M/e_n) = \frac{\displaystyle\sum_{w=m}^{k} \binom{k-m}{w-m} \frac{p(H_w)}{\pi(n, w\rho)} \cdot \frac{n+m\rho}{n+w\rho}}{\displaystyle\sum_{v=m}^{k} \binom{k-m}{v-m} \frac{p(H_v)}{\pi(n, v\rho)}}$$

and this depends clearly only on $n$ and $m$, which means that NH1 is satisfied,  Q.E.D.

For completeness we add:

$$(42) \qquad p(\overline{M}/e_n) = \frac{\displaystyle\sum_{w=m+1}^{k} \binom{k-m}{w-m} \frac{p(H_w)}{\pi(n, w\rho)} \cdot \frac{(w-m)\rho}{n+w\rho}}{\displaystyle\sum_{v=m}^{k} \binom{k-m}{v-m} \frac{p(H_v)}{\pi(n, v\rho)}}.$$

The proof of the other part of the equivalence:

(T12.2)   An NH-system is an SH-system

is much more complicated. A possible way to try to prove T12.2 is of course to show that an NH-system satisfies the axioms H1–H4. In our attempt to do so there were no essential difficulties in proving H2–H4. Moreover we could prove that $p(Q_i/H_w e_n)$ is a function depending at most on $w$, $n$, $n_i$ and $m$. However, we had to give up the attempt to show the final step leading to H1: that $p(Q_i/H_w e_n)$ does not depend on $m$.

Fortunately, it turned out to be possible to prove the theorem in a completely different way. The main idea behind this proof is as follows. In the context of T3 we remarked already that a choice of the $h(m, m)$-parameters in accordance with (8) does not guarantee that they give rise, together with $\rho$, to an adequate NH-system, i.e., to a probabilistic pattern satisfying the NH-principles. Now we start with a particular choice of the parameters $h(m, m)$ and $\rho$, and try to construct an SH-system in accordance with these parameters. If this succeeds we may conclude that the parameters give rise to an NH-system which is equivalent to the constructed SH-system, for we know from T12.1 that the latter system is also an NH-system and T3 excludes the possibility that there are two different NH-systems with corresponding parameter-values. If we do not succeed in constructing such an SH-system we shall show that there is something wrong with the parameters: they do not give rise to a probabilistic pattern satisfying the NH-principles.

That this line of attack could be successful, if the theorem is correct, was suggested by the comparison of T3 and the restricted version of T4 for SH-systems. For they say that both types of systems are characterized by the same number of parameters. From (36) and (9) it follows directly that a particular NH-system, with a particular value of $\rho$, can only be an SH-system with the same value for the 'SH-parameter' $\rho$. Apart from $\rho$ an NH-system is determined by the $k - 1$ parameters $h(m, m)$, $m = 1, 2, \ldots, k - 1$ and an SH-system by the $k - 1$ parameters $p(H_w)$, $w = 1, 2, \ldots k - 1$, or, equivalently, the $k - 1$ parameters $p(H_w)$, $w = 2, 3, \ldots, k$. Hence, if the theorem is correct, in both ways, one set of parameters need to be definable in the other and vice versa.

$^1$*Proof* of, T12.2. We start with a particular set of parameters $h(m, m)$, $m = 1, 2, \ldots, k - 1$ satisfying (8) and a finite value for $\rho$. The proof of the case that $\rho$ is infinite is essentially the same, but most formulas then get a different form. Let $p$ be the NH-system to which the parameters are supposed to give rise. For $m = 1, 2, \ldots, k$ holds

$$(43) \qquad p(e_m) = \frac{h(0, 0)}{k} \cdot \frac{h(1, 1)}{k - 1} \cdots \frac{h(m - 1, m - 1)}{k - (m - 1)}$$

$$= \frac{(k - m)!}{k!} \cdot \prod_{s=0}^{m-1} h(s, s) \qquad (e_m \in K^m, |M(e_m)| = m).$$

Since we are trying to construct an SH-system, $p'$, completely corresponding to the purported NH-system, there need to be values $p'(H_w)$, $w = 1, 2, \ldots, k$ such that we have, in particular, $p(e_m) \equiv p'(e_m)$. We have already argued that the $\rho$-parameters have to be the same. Hence, in the light of (43) and (39), our program leads to the following set of $k$

equations with $k$ unknowns $p'(H_w)$:

$$E(m) \qquad \sum_{w=m}^{k} \binom{k-m}{w-m} p'(H_w) \frac{\rho^m}{\pi(m, w\rho)} = \frac{(k-m)!}{k!} \cdot \prod_{s=0}^{m-1} h(s, s)$$

$$m = 1, 2, \ldots, k.$$

The set of equations has a unique solution as soon as they are mutually independent. This is, however, easy to see, for $E(k)$ has only one unknown, $p'(H_k)$, and $E(m)$ has, for $m < k$, one unknown more than $E(m + 1)$, viz., $p'(H_m)$. Because $h(0, 0) = 1$ it follows from $E(1)$ that this unique solution is such that

$$(44) \qquad \sum_{w=1}^{k} \binom{k}{w} p'(H_w) = 1.$$

Because of our assumption that the parameters $h(m, m)$, $m = 1, 2, \ldots$, $k - 1$, are all positive we may conclude from $E(k)$

$$(45) \qquad p'(H_k) > 0$$

(a)      Suppose now that the rest of the solution is non-negative, i.e.,

$$(46) \qquad p'(H_w) \geq 0, \quad w = 1, 2, \ldots, k - 1.$$

If we compare (44), (45) and (46) with (14) and T4 we see that $p'(H_w)$, $w = 1, 2, \ldots, k - 1$, together with $\rho$, are appropriate to determine a unique SH-system. Of course we indicate that SH-system by $p'$. By T12.1 this SH-system is also an NH-system. From our point of departure, $p(e_m) = p'(e_m)$ follows directly that $p'(\overline{M}/e_m) = p(\overline{M}/e_m) = h(m, m)$, $m = 1, 2, \ldots, k - 1$. By T3 the existence of two different NH-systems (with the same $\rho$) for which this holds, is excluded. Hence if (46) holds, then $p(Q_i/e_n)$ is uniformly equal to $p'(Q_i/e_n)$ and therefore we may conclude that $p$ is an SH-system.

(b) Now we set aside the assumption (46), i.e., that the solution is non-negative. The following definition is suggested by (42), the expression for $p(\overline{M}/e_n)$ in an SH-system:

$$(47) \qquad h'(n, m) =_{df} \frac{\displaystyle\sum_{w=m+1}^{k} \binom{k-m}{w-m} \frac{p'(H_w)}{\pi(n, w\rho)} \cdot \frac{(w-m)\rho}{n+w\rho}}{\displaystyle\sum_{v=m}^{k} \binom{k-m}{v-m} \frac{p'(H_v)}{\pi(n, v\rho)}},$$

$$1 \leq m \leq \min(n, k - 1),$$

provided that the denominator is non-zero. Note that $g'(n, m) =_{\text{df}} 1 - h'(n, m)$ is similar to (41). It is easy to check that the equations $E(m)$ guarantee that the denominator in (47) for $n = m$ is non-zero, because $0 < h(m, m) < 1$ for $m = 1, \ldots, k - 1$. Moreover, they imply that

(48)    $h'(m, m) = h(m, m),\quad m = 1, \ldots, k - 1.$

Suppose now that the denominator in (47) is non-zero for all $n$ and $m$ ($1 \le m \le \min (k - 1)$. Write the double recursive equation (10) for NH-systems as a general equation:

(10)'    $(1 - y(n, m)) y(n + 1, m) = y(n, m)(1 - y(n + 1, m + 1))$ .

$$\cdot \frac{n + m\rho}{n + 1 + (m + 1)\rho}$$

with $n = 1, 2, \ldots; 1 \le m \le \min (n, k); y(n, k) = 0, n \ge k$. Now it is possible to show that $y(n, m) = h'(n, m)$ is a solution of (10)' (*Hint:* substitute for $(1 - y(n, m))$ and $(1 - y(n + 1, m + 1))$ the formulas for $g'(n, m)$ and $g'(n + 1, m + 1)$ respectively which are similar to (41).) From this, (48) and T3 it follows that (47) is the explicit solution of (10) for our NH-system:

(49)    $h(n, m) = h'(n, m),\quad m = 1, 2, \ldots, \min (n, k - 1).$

(Note that this is not surprising at all, for in the case of the non-negative solution we know already that the NH-system is an SH-system and this implies that $p(\overline{M}/e_n)(= h(n, m))$ is uniformly equal to $p'(\overline{M}/e_n)$. The latter values are of course given by (47) and the former have to satisfy the recursive equation (10).)

Now we turn to the case that the denominator in (47) is not always non-zero. Let $m = m_0 < k$ be the largest $m$ and, for this $m, n = n_0 + 1$ the smallest $n$ for which this happens. Note that $n_0 + 1 > m_0$. It is easy to check now that $g'(n_0, m_0) = 0$ and that the proof of (49) remains valid for all $n$ and $m$ for which either $m_0 < m \le \min (n, k - 1)$ or $m = m_0$ and $m_0 \le n \le n_0$. But this implies that $h(n_0, m_0) = 1$, which is in conflict with (8). Hence, for a genuine NH-system the equations $E(m)$ are such that $h'(n, m)$ is always defined by (47).

(c) Suppose now that the solution of the equations $E(m)$ is such that $h'(n, m)$ is always defined by (47) but that it does not satisfy (46). That is, let $p'(H_w) < 0$ for some $w < k$. Let $u$ be the largest index for which this holds. It follows from (47) that the numerator of $h'(n, u)$ is positive for all

$n \geq u$. Its denominator becomes negative as soon as

$$\sum_{v=u-1}^{k} \binom{k-u}{v-u} p'(H_v) \frac{\pi(n, u\rho)}{\pi(n, v\rho)} < -p'(H_u).$$

Because $-p'(H_u) > 0$ this inequality holds if $n$ is large enough, for $\pi(n, u\rho)/\pi(n, v\rho)$ approaches 0 for $v > u$ (the proof is very similar to that of T3 of Chapter 5, Section 4.2) and therefore the whole left-hand sum approaches 0 by increasing $n$. We may conclude of course that, as soon as this happens, $h'(n, u)$ is negative and by (49) this is in conflict with (8). Therefore, our apparent NH-system is not a probability pattern and therefore also not a genuine NH-system,   *Q.E.D.*

From now on we shall call NH- or SH-systems with $p(H_k) < 1$: *Universalized Carnapian systems* (UC-systems). From T4 we know already that an SH-system with $p(H_k) = 1$ is in fact a Carnapian system (C-system). From the equivalence theorem we may conclude now that an NH-system is either a UC-system or a C-system. From the proof of that theorem we may conclude that an NH-system is a C-system if and only if $h(m, m) = (k - m)\rho/(m + k\rho)$ for $m = 1, 2, \ldots, k - 1$.


## 7. ANALYSIS OF UC-SYSTEMS

### 7.1. *General*
In this sub-section we shall analyze the Universalized Carnapian systems (UC-systems). The equivalence theorem of the preceding chapter enables us to use symbolizations for UC-systems based on the NH- as well as based on the SH-formulation. In the context of a particular UC-system $p$ we shall call the Carnapian system with ($w = k$ and) $\lambda = \rho k$ the *corresponding* C-systems $p_c$. In that system we have

(50)     $p_c(Q_i/e_n) = (n_i + \rho)/(n + k\rho)$.

The following theorem specifies the qualitative differences between UC-systems and their corresponding C-systems:

(T13.1)   For UC-systems holds in general

(a)      $p(\overline{M}/e_n) < (k - m)\rho/(n + k\rho) = p_c(\overline{M}/e_n)$,
         $m \leq \min(n, k - 1)$

(b)      $p(M/e_n) > (n + m\rho)/(n + k\rho) = p_c(M/e_n)$,
         $m \leq \min(n, k - 1)$

(c)    $p(Q_i/e_n) < \rho/(n + k\rho) = p_c(Q_i/e_n)$, $Q_i \notin M$

(d)    $p(Q_i/e_n) > (n_i + \rho)/(n + k\rho) = p_c(Q_i/e_n)$,    $Q_i \in M$.
All these inequalities become equalities for $m = k$ (clause (c) is in this case, of course, vacuous).

(T13.2)    For an SH-system with $p(H_k) = 1$ (and therefore $p(H_w) = 0$ for $w = 1, 2, \ldots, k - 1$) (i.e., an improper UC-system) holds $p(Q_i/e_n) \equiv p_c(Q_i/e_n)$: such systems are in fact C-systems.

*Proof:* The first inequality follows directly from (42) and the fact that $(w - m)\rho/(n + w\rho) < (k - m)\rho/(n + k\rho)$ for $w < k$. The other inequalities are trivial consequences of (a) combined with (36) and (37), which hold for both systems. From (41) it follows directly that $p(M/e_n) = 1$ for $m = k$ (in agreement with $g(n, k) = 1$ for $n \geq k$ in the NH-formulations); hence (b) transforms into an equality. It follows now immediately that this is also the case for the other ones. The claim in part 2 follows directly if one substitutes the extreme parameter-values in (41) and (42) and (36) and (37), *q.e.d.*

Notice that the inequalities do not hold generally for H-systems. Hintikka and Niiniluoto [19] have paraphrased the inequality-properties, in particular the last one, as: a UC-system is *more optimistic* than the corresponding C-system. This terminology makes, of course, only sense because the whole theorem can be interpreted as saying that, and specifying the way in which, a particular C-system is the extreme case of the corresponding class of UC-systems.

It might be thought that the requirement that the NH-parameters $h(m, m)$ are not larger than their corresponding C-values $((k - m)\rho/(m + k\rho))$ suffices to guarantee that all $h(n, m)$ are not larger than their corresponding C-values and therefore that such parameters give rise to a genuine NH-system. This is, however, not the case. The equivalence theorem enables us to add to T3 of Section 2: *For a particular value of $\rho$ the admissible combinations of the parameters $h(m, m)$ are determined by the requirement that the equations $E(m)$ should lead to a non-negative solution.* It is easy to see that this requirement implies that they may not be larger than their corresponding C-values.

It is nevertheless possible to give an explicit sufficient-but-not-necessary condition which guarantees that the equations have a *positive* solution (i.e., $p'(H_w) > 0$, $w = 1, 2, \ldots, k$) and therefore that we obtain a genuine NH- (or UC-) system.

(T14)    The condition

(51)        $0 < h(m,m) \leq \rho/(m + (m + 1)\rho), \quad m = 1, 2, \ldots, k - 1$

guarantees a positive solution of the equations $E(m)$.

*Proof:* We know already (45) that $p'(H_k) > 0$. Now, from $E(m)$ and $E(m + 1)$, $m < k$, we may derive

$$p'(H_m) = \sum_{w=m+1}^{k} \binom{k-m}{w-m} p'(H_w) \cdot \frac{\pi(m, m\rho)}{\pi(m, w\rho)} \cdot \left( \frac{(w-m)\rho}{(m + w\rho)h(m, m)} - 1 \right).$$

It is easy to check now that $p'(H_m) > 0$ if the condition is combined with the inductive hypothesis that

$$p'(H_w) > 0, \quad w = m + 1, m + 2, \ldots, k, \quad q.e.d.$$

From Section 4 we may of course conclude that UC-systems have the properties of instantial confirmation (29), universal confirmation (28) and universal-instance confirmation (30) as well as the properties of instantial, universal and universal-instance convergence (T8–T10).

UC-systems, however, have also some interesting inequality-properties which do not hold generally for H-systems. The most important ones will be stated in the following two theorems:

(T15)    strong universal confirmation (for UC-systems)

(52)        $p(H_M/e_n Q_i) > p(H_M/e_n), \quad Q_i \in M, M \neq K, p(H_M) > 0.$

This property can easily be proved on the basis of (28), (35.1) and (36). Together with T13 it justifies our proposal to talk about Universalized Carnapian systems.

(T16)    strong universal-instance confirmation (for UC-systems)

(53)        $p(M/e_n Q_i) > p(M/e_n), \quad Q_i \in M, M \neq K$
            if $\rho_K = \infty$ only if $p(H_W) > 0$ for some $W \supset M, W \neq K.$

The proof of this theorem is similar to that of T6.
    From (36) it follows, moreover, for UC-systems:

(54)        $p(Q_i/e_n Q_i M) > p(Q_i/e_n M), \quad Q_i \in M, M \neq \{Q_i\}.$

Note that (53) and (54) imply together the property of instantial confirmation (29).

We conclude this survey of inequalities with

(55)    $p(Q_i/e_nQ_i) > p(Q_i/e_nQ_j)$,    $Q_i, Q_j \in M$.

This follows directly from (36) and the fact that in all UC-systems holds (by definition in the NH-formulation) $p(M/e_nQ_i) = p(M/e_nQ_j)$ for $Q_i$ and $Q_j$ in $M$.

### 7.2 *Structurally Indifferent UC-systems: UC\*-systems* ($\rho = 1$)

In this sub-section we pay attention to UC-systems with $\rho = 1$: UC\*-systems. Remember that $\ddot{e}_n$ is the subset of $K^n$ containing those $e'_n$ for which $n_i(e'_n) = n_i(e_n)$, i.e., all order-permutation of $e_n$. Using the results of Chapter 5, Section 4.4, we may conclude that we obtain a UC-system with $\rho = 1$ by replacing H2 by the, conditionally imposed, strong version of POI:

(56)    $qw(e_n) = qw(e'_n)$,    $e'_n \in \ddot{e}_n$; $e_n \in W^n$

and H1 by the, conditionally imposed, principle of structural indifference (PSI):

(57)    $qw(\ddot{e}_n) = qw(\ddot{e}'_n)$,    $e_n, e'_n \in W^n$

and, finally, of course H3 (H4 is superfluous).

The argument is as follows: (56) implies

(58)    $qw(\ddot{e}_n) = \dfrac{n!}{\prod_i n_i!} qw(e_n)$

and (57) implies

(59)    $qw(\ddot{e}_n) = 1 \Big/ \left( \begin{array}{c} n + w - 1 \\ w - 1 \end{array} \right)$.

On the basis of (58) and (59) we can calculate $qw(e_n)$ and, of course, also $qw(e_nQ_i)$. We obtain then for $qw(Q_i/e_n)$: $(n_i + 1)/(n + w)$; this corresponds clearly with (20) for $\rho = 1$.

Another way to introduce these UC-systems, the NH-approach so to speak, is as follows. Replace POI by its unconditional strong version:

(SPOI)    $p(e_n) = p(e'_n)$,    $e'_n \in \ddot{e}_n$

and replace WPERR by

(WPSI)  *weak principle of structural indifference*

$$p(\ddot{e}_n) = p(\ddot{e}'_n), \quad m(e'_n) = m(e_n).$$

From WPSI it follows that we may define $p(\ddot{e}_n) = b(n, m)$. SPOI implies, analoguous to (58):

(60)     $$p(\ddot{e}_n) = \frac{n!}{\prod\limits_i n_i!} p(e_n).$$

Therefore we may conclude that

(61)     $$p(e_n) = \frac{\prod\limits_i n_i!}{n!} b(n, m).$$

On the basis of (61) it is easy to check that $p$ satisfies all NH-principles, and therefore that $p$ is an NH-system, in such a way that the NH-parameter $\rho$ is 1.

From (39) and (60) we may conclude that in UC*-systems holds

(62)     $$p(\ddot{e}_n) = \sum_{w=m}^{k} \binom{k - m}{w - m} p(H_w) \Big/ \binom{n + w - 1}{w - 1},$$

and from (40) we get:

(63)     $$p(H_w/e_n) = \frac{\dfrac{(w - 1)!}{(n + w - 1)!} p(H_w)}{\sum\limits_{v=m}^{k} \binom{k - m}{v - m} \dfrac{(v - 1)!}{(n + v - 1)!} p(H_v)}.$$

7.3  *Extreme UC-systems:* $\rho = \infty, \rho = 0$

$\rho = \infty$.  The value $\infty$ for $\rho$ falls within our definition of UC-systems. But the formulas for this case have a different form. For completeness we give here the reformulated versions of the main equations ((33), (36), (38)–(41)).

(64)     $$p(Q_i/H_w e_n) = 1/w$$

(65)     $$p(e_n/H_w) = (1/w)^n$$

(66)     $$p(e_n) = \sum_{w=m}^{k} \binom{k - m}{w - m} p(H_w)(1/w)^n$$

(67)    $p(H_w/e_n) = p(H_w)(1/w)^n \sum\limits_{v=m}^{k} \binom{k-m}{v-m} p(H_v)(1/v)^n$

(68)    $p(Q_i/e_n M) = 1/m, \quad Q_i \in M$

(69)    $p(M/e_n) = \sum\limits_{w=m}^{k} \binom{k-m}{w-m} p(H_w)(1/w)^n (m/w)$

$\sum\limits_{v=m}^{k} \binom{k-m}{v-m} p(H_v)(1/v)^n.$

The C-system with $\rho = \infty$, in which $p_c(Q_i/e_n) = 1/k$, can be obtained from the principle

(70)    $p_c(e_n) = p_c(e_n').$

For (70) implies that $p_c(e_n)$ is the inverse of the number of different $e_n \in K^n$ and this number is of course $k^n$. Notice that (70) includes the strong version of POI.

UC-systems with $\rho = \infty$ can be obtained by weakening (70) to

(71)    $p(e_n) = p(e_n'), \quad m(e_n) = m(e_n').$

Note first that (71) also includes SPOI. It implies moreover that $p(e_n)$ depends only on $n$ and $m$. It is now easy to derive that $p$ satisfies the NH-principles such that (68) holds and therefore such that $\rho = \infty$.

It is also possible to obtain these systems in the H approach: we need H3 and the conditional version of (70):

(72)    $qw(e_n) = qw(e_n'), \quad e_n, e_n' \in W^n.$

Whereas C-systems with $\rho = \infty$ do not satisfy any of the confirmation and convergence properties, discussed in Section 4 and Subsection 7.1, UC-systems with $\rho = \infty$ do have the properties of universal and universal-instance confirmation (even in the strong sense) and convergence. Moreover they have the property of instantial confirmation as long as $m < k$. But they do not have the property of instantial convergence, for $p(Q_i/e_n)$ goes to $1/m$ if $n$ grows without limit and $m$ remains constant. All these claims are easy to check.

$\rho = 0$.    Although we have excluded $\rho = 0$ in the definitions of all systems discussed in this chapter, we may substitute $\rho = 0$ in order to see what happens. In the SH-framework we get $p(Q_i/e_n) = n_i/n$, i.e., the straight rule, defined as the C-system with $\rho = 0$. Hence if $n_i = 0$ then $p(Q_i/e_n) = 0$

and therefore the value $\rho = 0$ is excluded by the regularity assumption. Note that we also have $p(M/e_n) = 1 = p(H_m/e_n)$. Hence, the *a priori* values $p(H_w)$ do not play any *a posterior* role. In other words, all straight, improper, UC-systems are *a posterior* equivalent to the straight rule.

With respect to the confirmation and convergence properties the situation is as follows: straight UC-systems would have the confirmation properties if the definitions of these properties had been stated in terms of 'larger than *or equal to*', except the case of instantial confirmation for these systems have that property already in a remarkable simple way within the given definition, provided that $n_i < n$. Finally, the convergence properties are all satisfied, although in a completely trivial way.

We are not sure whether substitution of $\rho = 0$ in the NH-framework only gives rise to the above discussed straight UC-systems and not to other types of systems. In other words, we do not know whether '$h(n, m)$ is uniformly equal to 0 (except for $n = m = 0$)' is the only solution, with $0 \leq h(n, m) \leq 1$, of the recursive equation (10) with $\rho = 0$.

### 8. FUNDAMENTAL DISCUSSION RELATED TO APPLICATIONS

The NH-parameters differ from the SH-parameters in that the former are all defined on finite Cartesian products of $K$, whereas the latter are defined on the infinite Cartesian product of $K$. We say that the former are based on finite and the latter on infinite arguments, or, for short, that the former are *finite* parameters and the latter *infinite* ones. At first sight this difference also holds between the $\rho$-parameters in both systems, but it follows from the equivalence theorem that the SH-parameter $\rho$ can also be determined as a finite parameter (e.g., on the basis of $p(Q_i/Q_jQ_jQ_iM) = (1 + \rho)/(3 + 2\rho)$). In what follows we shall therefore restrict our discussion to the parameters $h(m, m)$ and $p(H_w)$ respectively.

It is also possible to distinguish principles based (only) on finite arguments and those based on (some) infinite ones, or, again for short, *finite* and *infinite* principles. In this terminology all NH-principles are finite and the H-principles are infinite. Of course, finite principles, such as POI and WPERR, relate infinitely many values attached to finite arguments.

Our problem is whether, in the context of applications, finite parameters and principles are in any fundamental sense preferable to infinite ones. This problem is of particular importance in the standard application-context of an interpreted language. In this context there is a denumerably infinite universe and $K$ is interpreted as a family of properties (i.e., mutually exclusive and together exhaustive) with respect to that universe. The outcome $e_n Q_i$ adds then to the outcome $e_n$ the information that the $(n + 1)$-th individual has turned out to have the property $Q_i$. In this context $H_W$ corresponds to the universal hypothesis, $C_W$, that precisely the properties in $W$ are instantiated in the universe.

Of course it is only worth considering the application of an H-system, as explication of rational degrees of belief in such a context, if one is willing to accept that some $H_W (w < k)$ gets a positive *a priori* value, and therefore the same for the corresponding $C_W$. In other words it has to be acceptable that $p(C_K) = p(H_K) < 1$, which may be circumscribed as: there should be some *a priori* optimism with respect to uniformity (uniformity-optimism).

If one is not willing to accept some uniformity-optimism, then the only remaining H-type system is a Carnapian system in which $p(H_K) = 1 = p(H_K/e_n)$. Now it is important to note that, in the present context, $C_K$ is not equivalent to the analytical universal statement that all individuals have one of the properties belonging to $K$ for it tells also that all properties belonging to $K$ are instantiated in the universe. Therefore $C_K$ is a universal statement with empirical content and, by consequence, $H_K$ is not *a priori* certain. We have emphasized this point because it is frequently said that a C-system assigns zero probability to all empirical or non-trivial generalizations.

Many authors (e.g., Popper, Carnap and Essler) have expressed great doubts as to the acceptance of some uniformity-optimism and Popper has attempted to prove that it is impossible to do so in a consistent way. However all proofs suggested by Popper [31] have been shown to be invalid by Howson [20]. To be fair, Howson rejects uniformity-optimism too, but not on logical (mathematical) grounds but for epistemological reasons. In our opinion we may conclude from the fact that H-systems can be formulated in a completely consistent way that a general proof of the untenability of uniformity-optimism cannot be given. But then there remain the non-logical grounds, such as those of Howson [20] and Essler [9]. Here we shall only mention and not elaborate our hypothesis in this respect: all non-logical grounds against uniformity-optimism are different ways of articulating that our intuitions are against this type of optimism. At least in the present type of context we used to share these intuitions.

But let us try to disentangle possible objections to uniformity-optimism as far as the application of an H-system (not being a C-system) is concerned. Of course, either the H-principles or the H-parameters have to be objectionable.

The fundamental importance of the equivalence theorem now is that it reduces, in the case of SH-systems, this problem to the possible objection-ability of the finite NH-principles and NH-parameters. As for the NH-parameters $h(m, m)$ we do not see any objection in choosing them smaller than the corresponding Carnapian values (based on a previously chosen value for $\rho$). Of course, doing so is also a form of uniformity-optimism but now only with respect to singular outcomes. This type of optimism is comparable with the property of instantial confirmation shared by Carnapian systems ($\rho < \infty$). The basic point, for the present discussion, however, seems to be that we do not have intuitive objections to finite parameters as such. If one shares this intuitive feeling then there is also no objection to the parameters of an H-system not being an SH-system as soon as they can all be defined in terms of finite ones, which was already proved for SH-systems. In the next section we shall show that this is generally possible for H-systems.

There remains the question whether the finite NH-principles WPERR and POI are objectionable. If we come to the conclusion that they are not, then it does not follow that the H-principles H1, H2 and H3 are acceptable. For it might be that the systems which are excluded by the addition of H4, viz., the H-systems which are not SH-systems, are objectionable just because they violate H4. A possible strategy to tackle this problem would be to try to weaken WPERR in such a way that it gives rise to H-systems (together with POI) but remains finite. Unfortunately we did not succeed in finding such a principle (see Chapter 8, Section 2). Our research into this problem brought us also to the question of what type of (regular consistent probability) patterns satisfy: $p(e_n)$ is a real-valued function of the $n_i$'s. This principle might be called: equal relevance (PER). Of course PER implies POI and all H-systems satisfy PER. In other words the question is whether there are PER-systems not being H-systems. In Chapter 8, Section 1 we pay some attention to this question.

Now, are POI and WPERR objectionable? It is clear that POI is acceptable in all those application-contexts in which there is no reason to assume that the order of outcomes gives relevant information and this may certainly be supposed in the standard application-context. If $n_i(e_n) = n_i(e_n')$, then WPERR permits us to make a difference between $p(Q_i/e_n)$ and $p(Q_i/e_n')$ as soon as $m(e_n) \neq m(e_n')$. (PERR does not allow this.) In the present application-context this means that $p(Q_i/e_n)$ may depend on the variety observed in the sample of $n$ individuals. It seems very reasonable to leave room for such a dependency.

So we must conclude that we do not succeed in tracing back the objection to uniformity-optimism to some other objectionable feature of SH-systems. In our opinion we have to conclude that our intuitions are wrong: the resistance to accept uniformity-optimism is misplaced. We shall try to give an additional argument for this conclusion. Consider the following application-context. $K$ is a set of colours. Each point of a homogeneous ball has one of these colours. Suppose someone performs successive random throws with the ball and let us call the colour of the rest-point the outcome of a throw. Of course, the objective probability of a particular colour is proportional to the relevant surface and therefore constant. For simplicity we assume also that if a colour occurs on the ball, then its objective probability is positive. Let this be all the information we have at the start and let us come to know in addition the successive outcomes.

In this context $e_n$ is, of course, the (ordered) outcome of the first $n$ trials. $H_W$ corresponds to the hypothesis, $C_W$, that the colours on the ball are precisely those in $W$ in the following sense: if $C_W$ is true then the objective probability of $H_W$ is 1 and that of $H_V$ is 0 for all $V \neq W$. In other words, $p(H_W)$ is in fact our *a priori* degree of belief in $C_W$. The fundamental question is now whether we are inclined to start (and continue) with $p(C_K) = p(H_K) = 1$ or not, i.e., do we start to hold for certain that $C_K$ (all colours occur on the ball) is true, and therefore, also, do we continue to do so if we get overwhelming evidence compatible with an hypothesis to the effect that some but not all colours are on the ball. In our opinion it is highly unreasonable to do so and (at least) to our intuition we would, in fact, not do so. Hence, in this context, some uniformity-optimism seems required and commonly accepted.

Now, if one agrees with this analysis one may argue as follows. Although there are application-contexts in which our intuition is against uniformity-optimism, there are also contexts in which the contrary is the case. Apart from the probabilistic superstructures looked for, both types of contexts have the same set-theoretical structure and all certain information is the same. Therefore, it may well be that our intuitions in the first type of contexts are not sound and that we would do better in suppressing them and try to create the same intuitions as in the second type of contexts.

## 9. Finite parameters for H-systems

In the preceding section we have already said that our attempt to find a weakened (finite) version of WPERR such that it, together with POI, gives rise to the H-systems, was not successful. But there we have also argued that there is some relevance in a positive answer to the question whether there are finite parameters in terms of which the standard H-parameters can be defined.

Consider the following equations, holding in an H-system, for $m = 1$, $2, \ldots, k$,

$$I(m) \qquad p(e_m) = \sum_{w=m}^{k} \binom{k-m}{w-m} p(H_w) \frac{\rho_w{}^m}{\pi(m, w\rho_w)}$$

and for $Q_i \in M$ and $m = 2, 3, \ldots, k$

$$II(m) \qquad p(e_m Q_i Q_i) = \sum_{w=m}^{k} \binom{k-m}{w-m} p(H_w) \frac{\rho_w{}^m}{\pi(m, w\rho_w)} \cdot \frac{1+\rho_w}{m+w\rho_w} \cdot \frac{2+\rho_w}{m+1+w\rho_w}.$$

Now we can choose as finite parameters $p(e_m)$, $m = 1, 2, \ldots, k-1$ and (for $Q_i \in M$) $p(e_m Q_i Q_i)$, $m = 2, 3, \ldots, k$. For $I(k)$ and $II(k)$ are two equations with two unknowns, viz., $p(H_k)$ and $\rho_k$. Moreover, if we have calculated $p(H_w)$ and $\rho_w$ for $w = m+1, \ldots, k$, $(m > 1)$, then $I(m)$ and $II(m)$ lead together to $p(H_m)$ and $\rho_m$. Finally, $I(1)$ guarantees that

$$\sum_{w=1}^{k} \binom{k}{w} p(H_w) = 1.$$

Of course we may not choose the new parameters arbitrarily: they have to be such that they give rise to non-negative values for all $p(H_w)$ as well as for all $\rho_w$ otherwise they do not give rise to a regular probability pattern, assuming already that the equations are solvable.

The proposed parameters are of course not the only possible finite ones. An alternative set can be derived directly from the previous one: $p(M/e_m)$, $m = 1, 2, \ldots, k-1$ and $p(Q_i Q_i/e_m)$, $m = 2, 3, \ldots, k(Q_i \in M)$.

In both cases the second type of parameters $(p(e_m Q_i Q_i), p(Q_i Q_i/e_m))$ has been chosen instead of $p(e_m Q_i)$ and $p(Q_i/e_m)$ respectively, because in that case the relevant equations turn out not to do their job, as is easy to check.

Notice that the number of parameters is $2k - 2$, provided that the equations are mutually independent. If they are dependent their number will be smaller (e.g., in the case of SH-systems) but so will the number of infinite parameters.

## 10. REFORMULATION OF H-SYSTEMS; $k \to \infty$

Let us define $Z_w(n) = \bigcup_{|W| = w} H_W(n)$ and $Z_w = \bigcup_{|W| = w} H_W$. In an H-system we have now

$$(73.1) \quad p(Z_w) = \binom{k}{w} p(H_w)$$

$$(73.2) \quad \sum_{w=1}^{k} p(Z_w) = 1$$

$$(74.1) \quad p(Z_w/e_n) = \binom{k - m}{w - m} p(H_w/e_n)$$

$$(74.2) \quad \sum_{w=m}^{k} p(Z_w/e_n) = 1.$$

For fixed $e_n$, define $e_n$ as the subset of $K^n$ containing all those $e'_n$ (including $e_n$) which can be obtained from $e_n$ by a permutation of $K$, i.e., $e'_n \in e_n$ if there is a 1–1-function from $M(e_n)$ onto $M(e'_n)$ such that $e'_n$ is the result of transforming $e_n$ according to that function. If, for example, $e_8 = Q_6 Q_3 Q_6 Q_4 Q_8 Q_3 Q_3 Q_1$, then $e'_8 = Q_2 Q_3 Q_2 Q_1 Q_6 Q_3 Q_3 Q_8$ belongs to $e_8$; the function being '1, 3, 4, 6, 8 → 8, 3, 1, 2, 6'.

It is easy to see that we now have

$$(75) \qquad \text{if } e'_n \in e_n, \text{ then } p(e'_n) = p(e_n)$$

$$(76) \qquad \text{if } e'_n \in e_n, \text{ then } p(H_w/e'_n) = p(H_w/e_n).$$

From (75) it follows that $p(e_n) = |e_n| \cdot p(e_n)$. The number $|e_n|$ is equal to $k!/(k - m)!$, for there are $\binom{k}{m}$ subsets of $K$ with $m$ elements and to each subset correspond $m!$ different 1–1-functions. Hence we have

$$(77) \qquad p(e_n) = \frac{k!}{(k - m)!} p(e_n).$$

From (74) and (76) we obtain

$$(78) \qquad \text{if } e'_n \in e_n, \text{ then } p(Z_w/e'_n) = p(Z_w/e_n)$$

and, therefore, using (75)

$$(79) \qquad p(Z_w/e_n) = p(Z_w/e_n)$$

One way to derive

$$(80) \qquad p(e_n/Z_w) = \frac{w!}{(w - m)!} p(e_n/H_w), \quad w \geq m$$

is as follows. Substitute (77) and (79) in

(81)     $p(e_n/Z_w) = p(e_n)p(Z_w/e_n)/p(Z_w)$

and subsequently (73.1) and (74.1); and finally use $p(e_n/H_w) = p(e_n)p(H_w/e_n)/p(H_w)$.

These results, especially (77), (79) and (80), suggest the following reformulation of an H-system. Let $\mathbf{K} = \{\mathbf{Q}_1, \mathbf{Q}_2, \ldots, \mathbf{Q}_k\}$ and let $\mathbf{K}(n)$ be the subset of $\mathbf{K}^n$ containing all so-called well-ordered $n$-tuples, that is $n$-tuples starting with $\mathbf{Q}_1$ such that an occurrence of $\mathbf{Q}_i$ is preceded by one or more occurrences of $\mathbf{Q}_j$ for all $j = 1, 2, \ldots, i - 1$ and not preceded by occurrences of any $\mathbf{Q}_l$, $l = i + 1, \ldots, k$.

Now there is a unique function $f$ mapping all members of $K^n$ into $\mathbf{K}(n)$: $f(e_n)$ is obtained from $e_n$ by replacing (for all $i$ for which $n_i > 0$) all occurrences of $Q_i$ by $\mathbf{Q}_{d(i)+1}$, in which $d(i)$ is the number of different members of $M(e_n)$ occurring (one or more times) before the first occurrence of $Q_i$. It is easy to see that $e_n$ contains precisely those $e'_n$ for which $f(e'_n) = f(e_n)$. For example, both listed members of $e_8$ are mapped on $\mathbf{Q}_1\mathbf{Q}_2\mathbf{Q}_1\mathbf{Q}_3\mathbf{Q}_4\mathbf{Q}_2\mathbf{Q}_2\mathbf{Q}_5$.

In what follows $\mathbf{e}_n$ will be an arbitrary well-ordered $n$-tuple. We shall state explicitly $\mathbf{e}_n = f(e_n)$ if $\mathbf{e}_n$ is supposed to come from $e_n$. The meaning of $\mathbf{n}_i(\mathbf{e}_n)$ and $m(\mathbf{e}_n)$ is obvious. Of course we have now $\sum_{i=1}^{m} \mathbf{n}_i(\mathbf{e}_n) = n$.

We define $\mathbf{Z}_w(n) = \{\mathbf{e}_n \in \mathbf{K}(n)/m(\mathbf{e}_n) = w\}$ and $\mathbf{Z}_w = \bigcup_{n=w}^{\infty} \mathbf{Z}_w(n)\mathbf{K}_w\mathbf{K}_w\mathbf{K}_w\ldots$, in which $\mathbf{K}_w = \{\mathbf{Q}_1, \mathbf{Q}_2, \ldots, \mathbf{Q}_w\}$. It is easy to see that we have now $f(\mathbf{Z}_w(n)) = \mathbf{Z}_w(n)$ and $f(\mathbf{Z}_w) = \mathbf{Z}_w$ and also $f^{-1}(\mathbf{Z}_w(n)) = Z_w(n)$ and $f^{-1}(\mathbf{Z}_w) = Z_w$. Given an H-system $p$, the transformation function gives rise to a probability pattern w.r.t. $\mathbf{K}(1), \mathbf{K}(2), \ldots$ such that, if $\mathbf{E}_n \subset \mathbf{K}(n)$ then $\mathbf{p}(\mathbf{E}_n) = p(f^{-1}(\mathbf{E}_n))$. Kolmogorov's extension theorem assures us that there is a unique extension to $\mathbf{K}(\infty)$ such that, if $\mathbf{E} \subset \mathbf{K}(\infty)$ (and measurable), then $\mathbf{p}(\mathbf{E}) = p(f^{-1}(\mathbf{E}))$. (Note that $\mathbf{E}$ is measurable only if $f^{-1}(\mathbf{E})$ is measurable.)

If $\mathbf{e}_n = f(e_n)$, then $f^{-1}(\mathbf{e}_n) = e_n$. Therefore $\mathbf{p}(\mathbf{e}_n) = p(e_n)$. Of course we have also $\mathbf{p}(\mathbf{Z}_w) = p(Z_w)$. Let $\mathbf{Z}_w \cap \mathbf{e}_n$ be the set of members of $\mathbf{Z}_w$ starting with $\mathbf{e}_n$, and $Z_w \cap e_n$ the set of members of $Z_w$ starting with an element of $e_n$. Then we have $f^{-1}(\mathbf{Z}_w \cap \mathbf{e}_n) = Z_w \cap e_n$. Therefore we have also $\mathbf{p}(\mathbf{e}_n/\mathbf{Z}_w) = p(e_n/Z_w)$. Now we can state the main properties of the reformulated patterns: H-systems. From (80) and (27) we obtain first:

(82) $\qquad p(e_n/Z_w) = \dfrac{w!}{(w - m)!} \displaystyle\prod_{i=1}^{m} \dfrac{\pi(\mathbf{n}_i, \rho_w)}{\pi(n, w\rho_w)},$

(83.1) $\qquad p(Q_i/Z_w e_n) = (\mathbf{n}_i + \rho_w)/(n + w\rho_w), \quad i = 1, 2, \ldots, m,$

(83.2) $\qquad p(Q_{m+1}/Z_w e_n) = (w - m)\rho_w/(n + w\rho_w).$

The main equations for the absolute pattern are

(84) $\qquad p(e_n) = \displaystyle\sum_{w=m}^{k} p(Z_w)p(e_n/Z_w)$

(85) $\qquad p(Z_w/e_n) = p(Z_w)p(e_n/Z_w)/p(e_n)$

(86) $\qquad p(Q_i/e_n) = \displaystyle\sum_{w=m}^{k} p(Z_w/e_n)p(Q_i/Z_w e_n), \quad i = 1, 2, \ldots, m + 1.$

We have constructed the H-systems on the basis of H-systems. Of course we may conclude now that an H-system is mathematically sound, even if the original set $K$ is not specified. In other words we may define an H-system $\mathbf{p}$ completely in terms of $\mathbf{K}$, with parameters $p(Z_w) \geq 0$, $w = 1$, $2, \ldots, k - 1$ such that $\displaystyle\sum_{w=1}^{k-1} p(Z_w) < 1$ and $0 < \rho_w = {}_{df}\lambda_w/w \leq \infty$, $w = 2, 3, \ldots, k$, such that (82)–(86) hold.

Now we shall investigate what happens if $k \to \infty$. If we take this limit in an H-system we get $p(H_w) = 0$ for there are then infinitely many subsets of size $w$. By consequence $p(e_n)$ is always 0. Therefore this limit-pattern is irregular and seems to collapse completely. But observe that $p(Z_w)$ need not be 0. This opens a perspective to formulate an H-system on the basis of $\mathbf{K}_\infty = \{Q_1, Q_2, Q_3, \ldots\}$.

In order to include $Z_\infty$, i.e., in terms of the original framework: $f\left((K_\infty)^\infty - \displaystyle\bigcup_{w=1}^{\infty} Z_w\right)$, we note first that there is a direct way to extend a C-system to $K_\infty$. Let $0 < \lambda_\infty \leq \infty$ and $p(Q_i/e_n) = n_i/(n + \lambda_\infty)$, $n_i > 0$, and $p(\overline{M}/e_n) = \lambda_\infty/(n + \lambda_\infty)$. In the present context we define therefore $p(Q_i/Z_\infty e_n) = \mathbf{n}_i/(n + \lambda_\infty)$, $p(Q_{m+1}/Z_\infty e_n) = \lambda_\infty/(n + \lambda_\infty)$. By consequence we have $p(e_n/Z_\infty) = (\lambda_\infty)^m \displaystyle\prod_{i=1}^{m}(\mathbf{n}_i - 1)!/\pi(n,\lambda_\infty)$. Note that, if $\lambda_\infty = \infty$, then $p(e_n/Z_\infty) = 1$ if $m = n$, otherwise it is 0.

An $H_\infty$ is now defined in terms of the parameters $p(Z_w) \geq 0$, $w = 1$,

$2, \ldots, \sum_{w=1}^{\infty} \mathbf{p}(\mathbf{Z}_w) \leq 1$, and $0 < \lambda_w \leq \infty$, $w = 2, 3, \ldots, \infty$, such that (82)–(86) hold again, after the relevant replacements of parameters ($\rho_w = \lambda_w/w$) and including $w = \infty$ in the summations of (84) and (86).

The confirmation and convergence properties of H-systems hold of course also for **H**-systems, but in reformulated terms. For simplicity we assume $p(\mathbf{Z}_w) > 0$, for all (finite) $w$, and $m(\mathbf{e}_n) < k$.

(T17)(1)  $\mathbf{p}(\mathbf{Z}_m/\mathbf{e}_n\mathbf{M}) > \mathbf{p}(\mathbf{Z}_m/\mathbf{e}_n)$

(2)      $\mathbf{p}(\mathbf{Q}_i/\mathbf{e}_n\mathbf{Q}_i) > \mathbf{p}(\mathbf{Q}_i/\mathbf{e}_n)$,  $i = 1, 2, \ldots, m$

(3)      $\mathbf{p}(\mathbf{M}/\mathbf{e}_n\mathbf{M}) > \mathbf{p}(\mathbf{M}/\mathbf{e}_n)$

(4)      $\mathbf{p}(\mathbf{Z}_m/\mathbf{e}_n) \rightarrow 1$   if $n \rightarrow \infty$ and $m$ remains constant

(5)      $\mathbf{p}(\mathbf{M}/\mathbf{e}_n) \rightarrow 1$   if $n \rightarrow \infty$ and $m$ remains constant

(6)      $\mathbf{p}(\mathbf{Q}_i/\mathbf{e}_n) \rightarrow \mathbf{n}_i/n$  if $n \rightarrow \infty$ and $\lambda_w/w = \rho_w < \infty$ for all (finite) $w$.

It is easy to check that $\mathbf{H}_\infty$-systems have all these properties too.

UC-systems (including $\mathbf{UC}_\infty$-systems, in which holds that $\lambda_\infty = \infty$) have, in addition, the confirmation properties of UC-systems in reformulated terms:

(T17)(7)  $\mathbf{p}(\mathbf{Z}_m/\mathbf{e}_n\mathbf{Q}_i) > \mathbf{p}(\mathbf{Z}_m/\mathbf{e}_n)$,  $i = 1, 2, \ldots, m$

(8)      $\mathbf{p}(\mathbf{M}/\mathbf{e}_n\mathbf{Q}_i) > \mathbf{p}(\mathbf{M}/\mathbf{e}_n)$,   $i = 1, 2, \ldots, m$.

## 11. GH-SYSTEMS AND GUC-SYSTEMS

The conditional patterns of H-systems are Carnapian systems. Instead of these we shall now use Generalized Carnapian systems (GC-systems). Consider, in addition to the requirement that the conditional patterns should be regular consistent probability patterns, the principles, for all $W \subset K, |W| > 1$.

(GH1)  $q_W(Q_i/e_n) = f_i^W(n, n_i)$,  $Q_i \in W$

(GH2)  $q_W(Q_iQ_j/e_n) = q_W(Q_jQ_i/e_n)$,  $Q_i, Q_j \in W$.

From these two principles it follows, together with SPL in the case $|W| = 2$, that there are real numbers $\lambda_W, 0 < \lambda_W \leq \infty$, and $\gamma_i^W =_{df} q_W(Q_i)$, for all $Q_i \in W, \gamma_i^W > 0, \sum_{Q_i \in W} \gamma_i^W = 1$, such that

(87)    $q_W(Q_i/e_n) = (n_i + \gamma_i^W \lambda_W)/(n + \lambda_W), \quad Q_i \in W.$

In addition to these conditional patterns we need a probability function on the $H_W$'s($W \neq \cdot$ ):

(88)(1)   $q(H_W) \geq 0, \quad W \subset K$

(2)     $\sum_{W \subseteq K} q(H_W) = 1$

(3)     $q(H_K) > 0.$

Notice that it is not only not necessary to assume H3 ($q(H_W) = q(H_V)$ if $w = v$) but that it would even be rather strange in view of the fact that the conditional patterns may differ for subsets with the same size.

We shall call the systems obtained in this way from definition (18), Generalized Hintikka systems (GH-systems). If $q(H_K) = 1$ the system is of course a GC-system (with positive value for $\lambda_K$). Because (21) ($q_W(H_W) = 1$) remains valid we have again $p(Q_i/H_W e_n) = q_W(Q_i/e_n)$. It is easy to check that GH-systems have all confirmation and convergence properties of H-systems (see Section 4). Finally, it is easy to see that it is not possible to use the reformulation of H-systems of the preceding section (but this does not exclude the possibility that GH-systems can be extended to the case that $K$ is infinite).

The parameters may of course be related. One possibility seems very natural. Put $\gamma_i^K = \gamma_i$ and define $\gamma^W = \sum_{Q_i \in W} \gamma_i$. Suppose now that we have

(89)    $\gamma_i^W = \gamma_i/\gamma^W$   for all $W \subset K.$

In addition to (89) there may of course also be functional relations between $\gamma^W$, $\lambda_W$ and $p(H_W)$. Consider in particular

(90)    $\lambda_W/\gamma^W = \text{constant} = _{df}\lambda.$

Then we have

(91)(1)   $p(Q_i/H_W e_n) = (n_i + \gamma_i \lambda)/(n + \gamma^W \lambda)$

(2)     $p(Q_i/H_W e_n M) = (n_i + \gamma_i \lambda)/(n + \gamma^M \lambda), \quad Q_i \in M$

(3)     $p(Q_i/H_W e_n \overline{M}) = \gamma_i/(\gamma^W - \gamma^M), \quad Q_i \notin M.$

Because the numerators of (91) do not depend on $W$ we may conclude that

$p(H_W/e_n)$ and $p(M/e_n)$ (and $p(\overline{M}/e_n)$) only depend on $n$ and $M$ (and the parameters). Therefore $p(Q_i/e_nM)$ depends only on $n$, $n_i$, $M$ and $Q_i$. In fact we have (as in UC-systems): $p(Q_i/e_nM) = p(Q_i/Hwe_nM)(Q_i \in W)$.

A GH-system satisfying (89), (90) and $p(H_K) < 1$ will be called a Generalized UC-system (GUC-system). Combining the way in which POI and WPERR gave rise to UC-systems and the way in which GC-systems were introduced we may conclude that a GUC-system can be obtained from POI and

(WPRR) *weak principle of restricted relevance*
$$p(Q_i/e_n) = f_{i, M}(n, n_i).$$

(For the proof of this claim see the hint at beginning of the Appendix.) Finally, GUC-systems have, as UC-systems, all confirmation and convergence properties, including the strong ones.

## 12. SURVEY OF SYSTEMS

Within the formalism of the preceding section it is easy to give a complete survey of the (regular) consistent probability patterns w.r.t. $K$, $K^2$, $K^3$, . . ., with $2 \leq |K| < \infty$, that have been treated in Chapters 5 and 6. The largest class is that of GH-systems, defined by (18) on the basis of (87), (88) and the constraints for the parameters $\rho_W = {}_{df}\lambda_W/w(0 < \rho_W \leq \infty)$ and $\gamma_i^W(\gamma_i^W > 0, \sum_{Q_i \in W} \gamma_i^W = 1)$. In all these systems (21)–(24) hold, which permits us to replace $q(H_W)$ by $p(H_W)$ and $q_W(e_n)$ by $p(e_n/H_W)$ or $p_W(e_n)$.

We shall indicate a class of systems as follows: e.g., UC indicates the class of UC-systems. It will be convenient to introduce GSH, the class Generalized Special Hintikka-systems, which is of course defined as the union of GUC and GC.

Scheme 1 shows the mutual relations of those subclasses of systems which have been studied more in particular. Scheme 2 specifies the conditions that are necessary and sufficient to obtain these subclasses.

For completeness we also give a survey of the fulfilment of the confirmation and convergence properties. *For simplicity we restrict this survey to cases where $\rho_K < \infty$ and, if $p(H_K) < 1$, then $p(H_W) > 0$ for all $W \subset K$.* All properties holding for GH-, GUC-, GSH- and GC-systems hold also, of course, for H-, UC-, SH- and C-systems respectively.

*Scheme 1:* mutual relations

$$(= \text{GUC} + \text{GC})$$

$$\text{GH} \qquad \text{>GSH} \qquad \text{>GC}$$

$$\vee \qquad\qquad \vee \qquad\qquad \downarrow$$

$$\text{H} \qquad \text{>SH} \qquad \text{>C}$$

$$(\text{UC} + \text{C} = \text{NH})$$

→ contains as special or extreme case

*Scheme 2:* necessary and sufficient conditions $(\varnothing \neq W \subset K, |W| = w)$

$p(H_K) < 1$ $\qquad\qquad\qquad\qquad\qquad\qquad\qquad p(H_K) = 1$

$\text{GH} - \text{GC}$ $\qquad\qquad \text{GUC} = \text{GSH} - \text{GC} \qquad\qquad \text{GC}$

$$\lambda_W = \lambda \gamma^W$$
$$\text{or } \rho_W = \frac{k}{w} \rho \gamma^W$$
$$\gamma_i{}^W = \gamma_i / \gamma^W$$

$\text{H} - \text{C}$ $\qquad\qquad \text{UC} = \text{SH} - \text{C} \qquad\qquad \text{C}$

$$\lambda_W = \lambda(w) \qquad\qquad \lambda_W = \frac{w}{k}\lambda$$
$$(\text{or } \rho_W = \rho(w)) \qquad (\text{or } \rho_W = \rho)$$
$$\gamma_i{}^W = 1/w \qquad\qquad\qquad\qquad \gamma_i = 1/k$$
$$p(H_W) = p(w)$$

$\varnothing \neq W \subset K; \quad |W| =_{\text{df}} w;$
$\gamma_i =_{\text{df}} \gamma_i^K; \qquad \gamma^W =_{\text{df}} \sum\limits_{Q_i \in W} \gamma_i$
$\lambda =_{\text{df}} \lambda_K; \qquad \rho_W =_{\text{df}} \lambda_W/w; \rho =_{\text{df}} \rho_K.$

*Confirmation properties*

—instantial confirmation                                    **GH**

$$p(Q_i/e_nQ_i) > p(Q_i/e_n)$$

—universal-instance confirmation                           **GH**

$$p(M/e_nM) > p(M/e_n) \quad M \neq K$$

—strong universal-instance confirmation          **GSH = GUC + GC**

$$p(M/e_nQ_i) > p(M/e_n) \quad M \neq K, Q_i \in K$$

—universal confirmation                                **GH − GC**

$$p(H_M/e_nM) > p(H_M/e_n) \quad M \neq K$$

—strong universal confirmation                    **GUC = GSH − GC**

$$p(H_M/e_nQ_i) > p(H_M/e_n) \quad M \neq K, Q_i \in K.$$

*Convergence properties* $(n \to \infty)$

—instantial convergence                                    **GH**

$$p(Q_i/e_n) \to n_i/n \qquad \text{if } \rho_W < \infty \text{ for}$$
$$\text{all } W \subset K$$

—universal-instance convergence                           **GH**

$$p(M/e_n) \to 1 \qquad \text{if } M \text{ remains constant}$$

—universal convergence                                **GH − GC**

$$p(H_M/e_n) \to 1 \qquad \text{if } M \text{ remains constant.}$$

Precise definitions of all these properties can be found in Section 4, except for the 'strong' ones; these have been defined in Section 7.1.

Interesting special cases of the foregoing subclasses of systems can be obtained in two ways: by fixing the values for $\lambda_W$ in a definite way and by fixing the values for $p(H_W)$. With respect to $\lambda_W$ we have in particular:

(*)$_r$      $\lambda_W = w$   (or $\rho_W = 1$)

(†)$_r$      $\lambda_W = \infty$   (or $\rho_W = \infty$)

(O)      $\lambda_W = 0$   (or $\rho_W = 0$).

Systems with these particular values for $\lambda_W$ might be indicated by the corresponding sign as superscript to the right, as has already been done in case of C*- and UC*-systems. Note that (O) is in fact not allowed: it leads to irregular (straight) systems. Note further that an H-system satisfying one of these conditions is in fact a UC- or C-system with that condition.

As to $p(H_W)$ we remark first that $p(H_K) = 1$ and $p(H_W) = 0$ for all $W \neq K$, is of course a special case of particular importance: GC- and C-systems. As to other special cases we list only distributions satisfying H3:

$$(*)_l \qquad p(H_W) = 1/k) \binom{k}{w}$$

$$(\dagger)_l \qquad p(H_W) = 1/(2^k - 1)$$

$$(\square) \qquad p(H_W) = \binom{k}{w} \bigg/ \left( \binom{2k}{k} - 1 \right).$$

Now we may put the relevant sign as a superscript to the left. In Chapter 7, Section 5 we shall come back to these and other distributions, but then in a particular paradigmatic context. There we shall also introduce the final axiom (H5) to obtain the so-called $\alpha$–$\lambda$-continuum as a special subclass of H-systems. In Chapter 8, Section 3 we shall pay some attention to the UC-system with $(*)_l$ and $(*)_r$: the $^l\text{UC}^{l*}$-system.

## APPENDIX TO SECTION 2 (PROOF OF T1)

The proof of T1 is highly analogous to the proof in the Appendix to Chapter 5 and even less complicated. That proof could have started with the proof of

$$(1) \qquad p(Q_i/e_nM) = (n_i + \gamma_i\lambda) \bigg/ \left( n + \lambda \cdot \sum_{Q_i \in M} \gamma_i \right).$$

T1 is obviously a special case of (1). It will not be difficult to generalize the following proof of T1 to a proof of (1).

(a) Remember that we may assume that $k > 2$ and that $p(Q_i/e_n)$ and, therefore, $p(Q_i/e_nM)(Q_i \in M)$ are positive. From POI it follows, by the product rule, that if $n_i > 0$, $n_j > 0$ (and $n_i + n_j \leq n - m + 2$ if $m > 2$ and $n_i + n_j = n$ if $m = 2$):

(2)      $p(M/e_n)p(Q_i/e_nM)p(M/e_nQ_i)p(Q_j/e_nQ_iM)$
         $= p(M/e_n)p(Q_j/e_nM)p(M/e_nQ_j)p(Q_i/e_nQ_jM)$

which may be transformed, on the basis of the NH-principles, into:

(3)      $k_m(n, n_i)k_m(n + 1, n_j) = k_m(n, n_j)k_m(n + 1, n_i).$

Of course, the probability axioms imply:

(4)      $\sum_{Q_i \in M} p(Q_i/e_nM) = 1.$

Using NH2.2 we obtain as special cases of (4):

(4.1)    $k_m(n, n - m + 1) + (m - 1)k_m(n, 1) = 1$
(4.2)    $k_m(n + 1, n - m + 1) + k_m(n + 1, 2)$
         $+ (m - 2)k_m(n + 1, 1) = 1.$
(b) Let $m > 2$; substitution of $n_j = 1$ in (3) gives, for $1 \le n_i \le n - m + 1$,

(5)      $k_m(n, n_i)k_m(n + 1, 1) = k_m(n, 1)k_m(n + 1, n_i).$

Substitution of $n_i = n - m + 1$ and $n_i = 2$ respectively in (5) leads to:

(5.1)    $k_m(n + 1, n - m + 1) = \dfrac{k_m(n, n - m + 1)}{k_m(n, 1)} \cdot k_m(n + 1, 1)$

(5.2)    $k_m(n + 1, 2) = \dfrac{k_m(n, 2)}{k_m(n, 1)} \cdot k_m(n + 1, 1).$

Note that the substitution $n_i = 2$ is only allowed generally because $m$ is supposed to be larger than 2.
     Substitution of (5.1) and (5.2) in (4.2) gives:

(6)      $k_m(n + 1, 1) = 1 \Big/ \left( \dfrac{k_m(n, n - m + 1)}{k_m(n, 1)} + \dfrac{k_m(n, 2)}{k_m(n, 1)} + m - 2 \right).$

With the following definition of $\rho_m$

(7)      $k_m(m + 1, 1) = {}_{df}(1 + \rho_m)/(m + 1 + m\rho_m)$

we are now in a position to prove:

(8)      for fixed $m > 2$ and $1 \le n_i \le n - m + 1$ holds
         $k_m(n, n_i) = (n_i + \rho_m)/(n + m\rho_m).$

*Initial step:* $n = m + 1$; hence $n_i$ is 1 or 2; $k_m(m + 1, 1)$ satisfies (8) by definition (7); that $k_m(m + 1, 2)$ satisfies (8) now follows directly from (4.1);

*Inductive step:* suppose (8) holds for fixed $n \geq m + 1$; it then follows from (6) that it also holds for $k_m(n + 1, 1)$ and from (5) that it holds for $k_m(n + 1, n_i)$, $1 < n_i \leq n - m + 1$;

*Final step:* that the claim is true for $n = m$, i.e., that $k_m(m, 1) = 1/m$, follows directly from (4) and NH2.2.

(c) Let $m \geq 2$; from POI, the product rule and the NH-principles it is easy to derive $p(Q_i\overline{M}/e_n) = p(\overline{M}Q_i/e_n)$, $(Q_i \in M)$, which gives

(9)     $g(n, m)k_m(n, n_i)h(n + 1, m) =$
        $h(n, m)g(n + 1, m + 1)k_{m+1}(n + 1, n_i)$.

Substitution of $m = 2$ in (9) using (8) for $m = 3$, leads to the conclusion that $k_2(n, n_i)(1 \leq n_i \leq n - 1)$ is of the form $f(n) \cdot (n_i + \rho_3)/(n + 1 + 3\rho_3)$. From (4) it follows that $k_2(n, n_i) + k_2(n, n - n_i) = 1$. This implies that $f(n) = (n + 1 + 3\rho_3)/(n + 2\rho_3)$, and therefore, with the definition $\rho_2 =_{df} \rho_3$, we may conclude that (8) holds also for $m = 2$ and $1 \leq n_i \leq n - 1$. Note that (8) holds trivially for $m = 1$.

(d) Finally, it follows from (9) that $k_m(n, n_i)/k_{m+1}(n + 1, n_i)$ may not depend on $n_i$. In combination with (8) (including now the case $m = 2$) this implies that $\rho_m$ has to be constant for all $m = 2, 3, \ldots, k$, say $\rho$, and therefore we have:

(10)     $k_m(n, n_i) = (n_i + \rho)/(n + m\rho)$.

The necessary and sufficient condition which guarantees that $k_m(n, n_i)$ is always positive is easily seen to be $-1 < \rho \leq \infty$. It is also easy to check that this condition assures that $k_m(n, n_i)$ is never larger than 1; in fact it is always smaller than 1, $\quad$ Q.E.D.

# CHAPTER 7

# RATIONAL EXPECTATION IN MULTINOMIAL
# CONTEXTS

## 1. CARNAP'S INTENDED APPLICATION

Carnap constructed the continuum of inductive methods (i.e., the set of C-systems, $0 < \lambda < \infty$) for application to the sentences of a first-order language, containing a finite family of $(Q-)$ predicates and a countable number of individual constants. Let $K$ be such a family of $Q$-predicates $Q_1, Q_2, \ldots, Q_k$. We consider first the case that there are only finitely many individual constants: $U_N = \{a_1, a_2, \ldots, a_N\}$. Let $U_n$ indicate the set containing the first $n$ individuals $a_1, a_2, \ldots, a_n$. A state-description of $U_n$ is a sentence of the form $Q_{i1}(a_1) \& Q_{i2}(a_2) \& \ldots \& Q_{in}(a_n)$ and will be indicated here by $e(U_n)$. The number of occurrences of $Q_i$ in $e(U_n)$ will be indicated by $n_i(e(U_n))$, or simply $n_i$.

Carnap's aim now was to assign logical probabilities (or degrees of confirmation or rational degrees of belief) to the different hypotheses $Q_i(a_{n+1})(i = 1, 2, \ldots, k)$ on the basis of a state-description of $U_n$: $f(Q_i(a_{n+1})/e(U_n))$. In a remarkable chain of reasoning [3] he came to the conclusion that the best assignment would be

(1)     $f(Q_i(a_{n+1})/e(U_n)) = (n_i + \lambda/k)/(n + \lambda), \quad 0 < \lambda < \infty,$

i.e., what we have called (the special values of) a C-system. According to Carnap the parameter $\lambda$ may or may not depend on $k$. On the other hand, it may not depend on $N$ because, in our terminology, Carnap required that $f$ was consistent with respect to extensions of $U_N$. On the basis of the theorems of probability theory it is now possible to obtain from (1) the probability values for all sentences, e.g., $f(e(U_n))$.

From our treatment of GC-systems in Chapter 5 it will be clear that Carnap's presentation is, from a mathematical point of view, more complicated than strictly necessary and this is due to the fact that Carnap presented the mathematical system and the intended application as a mixture. Of course we do not want to blame Carnap for this. Our retrospective remark serves only to make things clear.

Now let us have a closer look at Carnap's application. It is not yet an application in the sense described at the end of Section 2 of Chapter 5. It is still a formal one in the sense that the language is uninterpreted. But of course it is intended to be used in concrete situations, i.e., in cases in which the language is interpreted. In such a *material* application each individual constant refers to a unique individual and the $Q$-predicates refer to ($Q$-)properties for which it holds that each individual has precisely one of these properties. The set of individuals referred to by the individual constants is the universe (of discourse) of the interpreted language.

Now, a C-system satisfies POI:

(2)     $f(Q_i(a_{n+1}) \ \& \ Q_j(a_{n+2})/e(U_n)) = f(Q_j(a_{n+1}) \ \& \ Q_i(a_{n+2})/e(U_n))$.

In other words the enumeration of individuals is supposed to be uninformative w.r.t. to the distribution of $Q$-properties. In probability theory this assumption is usually formulated as: the enumeration is *random* with respect to the distribution of $Q$-properties. But this is another way of saying that the investigation of the universe is in fact based on successive random sampling without replacement.

The first question is, therefore, whether a C-system is acceptable as a rational probability pattern with respect to random sampling without replacement in a fixed universe. And, if the answer would be positive, the second question is whether the parameter $\lambda$ may or may not depend on $N$, apart from a possible dependency on $k$. In Section 6 we shall come back to the first question. As to the second question we think that $\lambda$ needs to depend on $N$

Consider the case that $K = K_3 = \{$white, black, red$\}$. Now, we are strongly inclined to assign a higher probability to the hypothesis that the tenth individual will be white, given that the first nine are white, in case $N = 10$ than in case $N = 10,000$. Our intuitions with respect to this example are certainly influenced by our knowledge of statistics, but we are not able to base conclusive arguments in favour of our intuitions on statistics. We conclude from this example that, if two contexts differ only with respect to $N$, then some special values, related to corresponding hypotheses and evidence, will differ and therefore, in general, that $\lambda$ needs to depend on $N$. Here we do not enter into the question of how the given example can be generalized to a qualitative condition of adequacy.

Carnap left room for the dependency on $k$ but not for the dependency on $N$. In our opinion, our example arguing for the dependency on $N$ leaves only two possibilities: either we have to accept that dependency or we have to reject the applicability of a C-system to random sampling without replacement and therefore to Carnap's intended (material) application. Of course we might come to the latter conclusion without rejecting the applicability of a C-system to the limit-case: a (denumerably) infinite universe.

## 2. THE MULTINOMIAL CONTEXT

In the preceding section we have treated Carnap's intended application without explicit reference to the general approach of rational expectation in a paradigmatic context as described in Chapter 4. In the following sections we shall do so. Moreover, in the line of the preceding two chapters, we shall avoid symbolic notations containing inessential information, such as the reference to particular individuals in (2).

In this section we shall present the context, which is, in our opinion, the most suited for the application of UC-systems: the open multinomial context. Let each experiment of a paradigmatic context be in fact the performance of a so-called repeatable experiment, i.e., an experiment with constant, but unknown, objective probabilities $q_1, q_2, \ldots, q_k$ for the elementary outcomes $Q_1, Q_2, \ldots, Q_k$. The objective probability pattern governing the outcomes of a repeatable experiment is given by

$$(3.1) \quad 0 \leq P(Q_i) = P(Q_i/e_n) = q_i \leq 1 \qquad (3.2) \quad \sum_{i=1}^{k} q_i = 1.$$

The product rule leads then, with $x^0 = 1$, to

$$(4.1) \quad P(e_n) = \prod_{i=1}^{k}(q_i)^{n_i} \qquad (4.2) \quad P(\ddot{e}_n) = \left( n! \ \prod_{i=1}^{k} n_i! \right) \cdot \prod_{i=1}^{k}(q_i)^{n_i}$$

where (4.2) is known as the multinomial distribution.

There is of course one subset of $K$ for which holds that all its members and no others have positive probability. Let $C_W$ indicate the hypothetical statement that $W(W \subset K, W \neq \varnothing)$ is this set: $C_W =_{\text{df}} \forall_{Q_i \in K}(q_i > 0 \leftrightarrow Q_i \in W)$. $C_W$ is called a constitutional theory (of size $|W| = w$) or, simply, a (w-)*constituent*. For each $w$, $1 \leq w \leq k$, there are of course $\binom{k}{w}$ different w-constituents. Their disjunction, stating that the number of positive $q_i$ is $w$, is called the structural theory or constituent-structure of size $w$ or, simply, (w-)*structure*. Later on we shall include $C_\phi$ for which the interpretation depends on the context. Without these there are $2^k - 1$ constituents and $k$ structures. Moreover, both sets of hypotheses are mutually exclusive and together exhaustive. Therefore we may choose one of them as our set of elementary theories.

An *open multinomial (or K-) context* is now defined as a paradigmatic context based on a repeatable experiment in which the constituents are the elementary theories (and, by consequence, in which the structures are the theories defined as the sets of constituents of a particular size). In what follows we shall always assume that we do not know more about the underlying probability process, except in case we assume explicitly some additional knowledge, as, for example, in the following definition.

If it is known that $C_W$ is the true constituent we speak of a $C_W$-*context* or, more generally, of a *closed multinomial context*. In this case it may of course not matter whether $C_W$ is considered as the only elementary theory or not.

An example of a $K$-context, near to Carnap's intended application, is that of random sampling in an infinite universe. $C_W$ is then the hypothesis that precisely the members of $W$ are exemplified if we may assume in addition that $q_i > 0$ if $Q_i$ is exemplified. (If we know that $C_W$ is true it turns into a $C_W$-context.) Notice that it does not matter whether the sampling is (said to be) with or without replacement. The term constituent has been introduced by Hintikka [16] for the described hypothesis with respect to a universe. Carnap [5] has suggested the term constituent-structure for the hypotheses that the number of $Q_i$ exemplified in a particular universe is $w$.

Random sampling *without* replacement in a finite universe is of course not a multinomial context because the probabilities do change. This type of context will be studied in the last section. On the other hand, random sampling *with* replacement in a finite universe is an example of a multinomial context. But in this example we have some additional information about the probabilities: if we know the size of the universe ($N$) we know also that the $q_i$'s can only have the values $0, 1/N, 2/N, \ldots, 1$; if we do not know the size we still know that the $q_i$'s are rational numbers. For these reasons we shall not pay further attention to this type of multinomial context.

From the mathematical point of view the example of random sampling in a partitioned infinite universe is equivalent to the following ball-model. The surface of a (homogeneous) ball is partitioned by the $k$ colours $Q_1, \ldots, Q_k$ and the outcomes of the experiments are the colours of the points at which the ball comes at rest after successive throws. In this example the system of the context is of course the coloured ball. In that of random sampling it is the partitioned universe of individual objects.

### 3. FORMALLY RATIONAL PATTERNS FOR OPEN
### MULTINOMIAL CONTEXTS

It will be convenient to repeat some definitions of the preceding chapter (Section 3) and to add one other. $W$ is always a non-empty subset of $K, |W| = w$.

(5) $\qquad M(e_n) = M = \{Q_i \in K / n_i(e_n) = n_i > 0\}$

(6) $\qquad H_W(n) = \{e_n \in K^n / M(e_n) = W\}.$

The subset $H_W(n)$ of $W^n$ is empty if and only if $w > n$. For fixed $n$ the $H_W(n)$ constitute a partition of $K^n$.

(7) $\qquad H_W = \bigcup_{n=w}^{\infty} H_W(n) WWWW. \ldots$

$H_W$ is a measurable subset of $W^\infty$; the $H_W$ constitute a partition of $K^\infty$.

The members of $K^\infty$ can be divided into two disjoint classes: infinite sequences in which all occurring $Q_i$ occur infinitely many times and infinite sequences in which some $Q_i$ is occurring, but only a finite number of times. Let $G_W$ indicate the intersection of $H_W$ and the former subset of $K^\infty$.

Now we return to our open multinomial context. If $C_W$ is true, we know, with objective probability 1, that all members of $W$ will occur infinitely many times, for the probability that a particular $Q_i \in W$ (with $q_i > 0$) will never (re) occur is always $\lim_{n \to \infty} (1 - q_i)^n = 0$. Therefore $G_W$ is the set of infinite sequences compatible with $C_W$. The $G_W$'s are mutually non-overlapping. Therefore, the constituents are decidable, though not by a finite number of experiments.

For finite sequences all members of $W^n$ are compatible with $C_W$, because to any finite sequence we can add an infinite one belonging to $G_W$.

From Section 6 of Chapter 4 we may conclude that a formally rational expectation pattern with respect to a $K$-context has, as the prior belief function, a probability function on the $C_W$'s and as $C_W$-prediction patterns, consistent probability patterns w.r.t. (the $C_W$-spaces) $W, W^2, \ldots$ which need to be closed:

(8) $\qquad f_W(G_W) = 1$ (and therefore $f_W(H_W) = 1$).

The absolute prediction pattern is obtained from

(9) $\qquad f(e_n) = \sum_{W \supset M(e_n)} f(C_W) f_W(e_n)$

which leads to a consistent probability pattern w.r.t. $K, K^2, \ldots$ for which holds

(10)    $f(G_W) = f(H_W)(= f(C_W))$.

The posterior belief function is obtained from

(11)    $f(C_W/e_n) = f(C_W)f_W(e_n)/f(e_n), \quad W \supset M$.

A $K$-context is (countably) decidable (see Section 7 of Chapter 4), for the $G_W$ are mutually non-overlapping. Hence we may conclude that a consistent probability pattern w.r.t. $K, K^2, \ldots$ satisfying (10) determines a unique formally rational expectation pattern by the definitions

(12)    $f(C_W) = f(H_W)$

(13)    $f_W(e_n) = f(e_n/H_W)(= {}_{df}f(H_W \cap e_nWW\ldots)/f(H_W)), \quad e_n \in W^n$.

The search for a rational conditional prediction pattern $f_W$ is of course equivalent to the search for a rational (absolute) prediction pattern w.r.t. the corresponding closed multinomial $C_W$-context.

From now on we shall generally speak about (rational) probabilities because all functions looked for are required to be probabilistic.

### 4. MATERIAL CONDITIONS OF ADEQUACY; UC-SYSTEMS AS EXPECTATION PATTERN FOR OPEN MULTINOMIAL CONTEXTS

It is now our task to propose additional material conditions of adequacy to determine those (formally rational) assignments of probabilities which are acceptable in view of the particular nature of a $K$-context. Our final conclusion in this section will be that $f(e_n)$ has to be a UC-system with $0 < \rho < \infty$ and $f(H_w)(= f(C_W)) > 0$ for all $w$. This implies of course that $f_W(e_n)$ is a C-system with $0 < \lambda_W = w\rho < \infty$. We shall postpone proposals for the prior belief function to the next section.

We shall try to argue our proposal step by step. Unfortunately, this leads to some technical duplications with the preceding two chapters. The main purpose of what follows is however to show a general methodology for the construction of rational expectation patterns at work.

We shall use the following four methodological principles, listed in order of priority:

(MPR)    *Regularity:* Leave room, as far as possible, for all possible perspectives, i.e., assign, if possible, non-zero probabilities.

(MPL)   *Learning from experience:* Try to learn from experience, i.e., try to approach the objective state of affairs.

(MPI)   *Indifference:* Make no differences as long as you cannot give reasons for it.

(MPS)   *Simplicity:* If a choice is left open, then choose one of the simplest alternatives.

A number of remarks have to be made about these principles. A principle will not always have a unique interpretation in a concrete situation. In many cases, however, such a problem of choice will be solved by the priority-order between the principles. If this does not work it is of course advisable to postpone a decision as long as you can.

An application of one principle (e.g., MPI) may be at the same time an application of another (e.g., MPS) and, conversely, the application of one (e.g., MPL) may presuppose the application of another (e.g., MPR).

MPR contains an if-possible-clause, for it may be impossible to apply it, for example, to assign non-zero probabilities to the elements of a non-denumerable set.

Finally, the principles of consistency (CA2) and certainty (CA3) might have been presented as applications of MPI (indifference w.r.t. perspective-extensions, and indifference w.r.t. certain (or w.r.t. impossible))perspectives, respectively). However, we prefer to speak only of (applications of) MPI if it is applied in the case of incompatible hypotheses and/or conditional assumptions.

In what follows A refers to the absolute prediction pattern, B to the belief function and C to the conditional patterns. In addition to the probabilistic nature of all functions we still have the following, in view of the definitions (9), (12) and (13), mutually equivalent,

(FCC)   *formal closure conditions*

(A)     $f(G_W) = f(H_W)$

(C)     $f_W(G_W) = 1.$

We interpret MPR in the following straightforward sense:

(PR)    *principles of regularity*

(A)     $f(Q_i/e_n) > 0$

(C)     $f_W(Q_i/e_n) > 0, \quad e_n Q_i \in W^{n+1}$

(B)      $f(C_W/e_n) > 0, \quad W \supset M.$

In our opinion PR-A&C are unproblematic. In Section 8 of Chapter 6 we have already discussed the acceptability of some uniformity-optimism. Below we shall find an additional argument for the general uniformity-optimism

(GUO)   $f(C_W) > 0$   for all $W$

implied by PR-B. Note that PR-B&C imply PR-A but the converse holds only in combination with GUO. Here, as in what follows, we have stated explicitly only those equivalent formulations and mutual relations which are either not too elementary or very important.

Our first applications of MPL are rather plausible:

(PCf)    *principles of confirmation*

(A)      *instantial confirmation*
         $f(Q_i/e_nQ_i) > f(Q_i/e_n)$

(C)      *conditional instantial confirmation*
         $f_W(Q_i/e_nQ_i) > f_W(Q_i/e_n), \quad e_nQ_i \in W^{n+1}.$

We shall consider below a number of additional confirmation principles. But first we require, again on the basis of MPL :

(PCv)    *principles of convergence* $(n \to \infty)$

(A)      *instantial convergence*
         $f(Q_i/e_n) \to n_i/n$

(C)      *conditional instantial convergence*
         $f_W(Q_i/e_n) - n_i/n, \quad e_nQ_i \in W^{n+1}.$

(B)      *universal convergence*
         $f(C_M/e_n) - 1.$

The formulation of these, and the following, convergence conditions can be made precise in the same way as has been done in Section 4 of Chapter 6. Again, PCv-B&C imply PCv-A. Conversely, PCv-A implies PCv-B only in combination with GUO and this is a strong argument in favour of GUO. Put in a different way: PCv-B presupposes GUO, i.e., in order to learn from experience we have to be regular w.r.t. the $C_W$.

PCv-A and PCv-C imply respectively:

(PCv-AM) *universal-instance convergence*
$$f(M/e_n) \to 1$$

(PCv-CM) *conditional universal-instance convergence*
$$f_W(M/e_n) \to 1, \quad M \subset W.$$

Consider now the following pairs of additional confirmation principles

(PCf-AMs) *strong universal-instance confirmation*
$$f(M/e_n Q_i) > f(M/e_n), \quad M \neq K, Q_i \in M$$

(PCf-Bs) *strong universal confirmation*
$$f(C_M/e_n Q_i) > f(C_M/e_n), \quad M \neq K, Q_i \in M.$$

and

(PCf-AM) *universal-instance confirmation*
$$f(M/e_n M) > f(M/e_n), \quad M \neq K$$

(PCf-B) *universal confirmation*
$$f(C_M/e_n M) > f(C_M/e_n), \quad M \neq K.$$

The former are called strong because they imply the latter, i.e., PCf-AMs implies PCf-AM and PCf-Bs implies PCf-B. Although we are intuitively inclined to consider, in the first place, the strong versions, we prefer at this moment to be more cautious and to choose the weak versions. Again, PCf-B&C imply PCf-A and the converse presupposes GUO.

The universal confirmation and convergence properties are of course strongly related to the closure conditions but we shall not work out this in detail and turn to MPI.

The assumption of a repeatable experiment implies that we know that the order of outcomes does not give information about the objective state of affairs. Moreover, we do not know anything about differences between the predicates in $K$. Therefore, we require that our system is a PER-system, i.e., a system satisfying:

(PER)   *principles of equal relevance*

(A)      $f(e_n)$ depends at most on $n$, $m$ and the $n_i$

(C)      $f_W(e_n)(e_n \in W^n)$ depends at most on $n$, $m$, $w$ and the $n_i$

(B)      $f(C_W/e_n)(M \subset W)$ depends at most on $n$, $m$, $w$ and the $n_i$.

In this case PER-B&C iff PER-A, holds unconditionally, which is easy to see, but complicated to prove in detail. PER permits us of course to replace all index-occurrences of $W$ and $M$ by $w$ and $m$, respectively. Note that PER implies some of the principles constituting an H-system, viz., H2 and H3 and part of H1.

Let us concentrate now on $f_W(Q_i/e_n)$. This conditional special value may depend on $n, m, w, n_i$ and the other $n_j (j \neq i)$. From the equivalence theorem between NH-systems and SH-systems (T12 of Chapter 6, Section 6, see also T13.2 of Chapter 6. Section 7.1) it may be concluded that, if we drop the dependency on the $n_j (j \neq i)$, we have also to drop the dependency on $m$ in order to obtain $f_W(H_W) = 1$. But for what reason should we choose for a dependency on the $n_j(i \neq j)$ instead of a simple dependency on their sum $(n - n_i)$. which is already implied by a dependency on $n$ and $n_i$. Moreover, in view of MPS, we may say that restriction to a dependency on $n$ and $n_i$ (and $w$) is certainly the simplest solution. Hence we impose

(PERR-C)   (conditional) principle of equal restricted relevance
$f_W(Q_i/e_n)$ depends at most on $n, n_i$ and $w$

which is H1 of H-systems.

In summary, we have now that $f(e_n)$ is an H-system or, equivalently, that $f(C_W)$ depends only on $w$ and that the $f_W(e_n)$ are C-systems. From Chapter 6 we know that all imposed conditions of regularity, confirmation and convergence are satisfied if and only if the parameters are such that $0 < \rho_w < \infty$ and $f(C_w) > 0$ for all $w$, $1 \leq w \leq k$. Moreover, (21) of Chapter 6 tells us that a C-system is at least closed in the weak sense, i.e., $f_W(H_W) = 1$. That it is closed in the proper sense $(f_W(G_W) = 1)$ follows from the fact that. even if $e_n \in H_W(n)$, $f_W(V^\infty/e_n) = 0 (V \neq W)$, for which the proof is very similar to that of T3 in Chapter 5, Section 4.2. The mentioned fact implies that, after each occurrence of a $Q_i \in W$, it is with probability 1 that $Q_i$ will reoccur sooner or later.

For the final step of arriving at the conclusion that $f(e_n)$ has to be a UC-system, i.e., $\rho_w = \rho$ for all $w$. we may of course use H4, which comes to:

(PERR-CM)   $f_W(Q_i/e_nM)$ depends at most on $n, m, n_i$   $(W \supset M, Q_i \subset M)$.

Note that this is a genuine restriction on PERR-C for the latter implies only that $f_W(Q_i/e_nM)$ depends at most on $n, m, n_i$ and $w$. It is easy to check that instead of PERR-CM we could also choose, for example,

(PERR-B)    $f(C_W/e_n)$ depends at most on $n$, $m$, $w$    $(W \supset M)$

which contains PER-B.

The equivalence theorem implies that we could have avoided the intermediate step PERR-C (and, therefore, of H-systems) by imposing:

(WPERR-A) $f(Q_i/e_n)$ depends at most on $n$, $m$, $n_i$.

For completeness' sake we remember that WPERR-A is equivalent to the NH-principles:

(NH1)        $f(M/e_n)$ depends at most on $n$ and $m$

(NH2.1)      $f(Q_i/e_n\overline{M})$ depends at most on $m$    $(Q_i \notin M)$

(NH2.2)      $f(Q_i/e_nM)$ depends at most on $n$, $m$, $n_i$    $(Q_i \in M)$.

From Chapter 6, Section 7 we know that UC-systems have the strong confirmation properties PCf-AMs and PCf-Bs. We are not sure whether the step from H-systems to UC-systems is implied by the requirement that these strong properties have to hold. In any case their satisfaction seems intuitively highly plausible.

Thus far we have based our argument in favour of UC-systems (among PER-systems) on additional applications of MPI. But, in the light of MPS, we think that anyone will agree that UC-systems are not only the simplest H-systems but also the simplest PER-systems satisfying all conditions of adequacy (in the strict sense) w.r.t. closure, regularity, confirmation and convergence (provided of course that GUO holds and that $0 < \rho < \infty$).

In the preceding discussion we have not mentioned the stages at which the special principle of linearity (SPL) was required. From the preceding two chapters we may conclude that our final conclusion requires SPL only in the case $|K| = k = 2$. Then it must be applied either to $f_2(Q_i/e_n)$ or, for $M(e_n) = K$, to $f(Q_i/e_nM)(= f(Q_i/e_n))$. In our opinion these applications, as well as the one required to conclude to H-systems in case $k > 2$ (viz., the first one mentioned above), are very plausible in the light of the other results: they may be seen as applications of MPS, and in a way, also of MPI.

In the terminology of the end of Chapter 2 we may restate our conclusions as follows: *open multinomial contexts are appropriate for inductive application of UC-systems, and closed multinomial contexts for C-systems.*

In the next section we shall pay attention to the prior belief function, also called constitutional distribution. With respect to the choice of the parameter $\rho$, we confine ourselves to the remark that $\rho = 1$ leads certainly to the simplest UC-systems: the so-called structually indifferent UC*-systems, treated in Chapter 6, Section 7.2. Among others it is shown there that we can arrive at these systems by adding to SPOI:

WPSI    *weak principle of structural indifference*
        $f(\ddot{e}_n)$ depends at most on $n$ and $m$

in which $\ddot{e}_n$ is the set of order permutations of $e_n$.

## 5. Constitutional Distributions for Open Multinomial Contexts

In this section we suppose that $f(e_n)$ is an H-system. For the prior belief function holds then $f(C_W) = f(C_V) = f(C_w)$ if $|V| = |W|$. Therefore we have

(14)    $f(S_w) = \begin{pmatrix} k \\ w \end{pmatrix} f(C_w)$

where (the structure) $S_w$ is the set, or the disjunction, of constituents of size $w$.

It will be convenient to introduce $C_\emptyset = S_\emptyset = C_0 = S_0$. We may interpret $C_0$, for example, as the hypothesis that the repeatable experiment does not work, for some reason or other.

Our task is to specify acceptable *constitutional distributions* such that

(15)    $f(C_0) \geq 0, \quad f(C_w) > 0 < f(S_w), \quad 1 \leq w \leq k$

(16)    $\sum_{w=0}^{k} \begin{pmatrix} k \\ w \end{pmatrix} f(C_w) = 1 \left( = \sum_{w=0}^{k} f(S_w) \right).$

In view of (14) it is also sufficient to specify the *structural distribution* $f(S_w)$.

If $f(C_0) > 0$ then we may of course consider as the proper distributions, for $1 \leq w \leq k$,

(17)    $f'(C_w) = {}_{df} f(C_w)/(1 - f(C_0)) \qquad f'(S_w) = {}_{df} f(S_w)/(1 - f(S_0)).$

There are at least two serious candidates for the application of MPI (and MPS):

(CI-B)    $f(C_w) = f(C_v)$

(SI-B)    $f(S_w) = f(S_v).$

CI-B leads to $f(C_w) = 2^{-k}$ and $f'(C_w) = 1/(2^k - 1)$. SI-B leads to $f(S_w) = 1/(k + 1)$ and $f'(S_w) = 1/k$. The SI-B distribution was in fact suggested by Carnap [5].

Either of these two alternatives could be based on the argument that the constituents and the structures respectively can be seen as (mutually exclusive and together exhaustive) properties of the context. Therefore, a choice between these alternatives seems rather arbitrary.

We have assumed to lack any knowledge about the $q_i$ of our multinomial context. This does not, however, exclude that our knowledge about the physical nature of the context suggests us some qualitative relations for the prior distribution, such as CI-B or SI-B, or their (common) weaker variant

(18)     $f(C_w) = f(C_{k-w})$,     $(\leftrightarrow f(S_w) = f(S_{k-w}))$.

Distributions satisfying (18) will be called *symmetric*. Qualitative relations leading to non-symmetric distributions are, e.g., for $w > v$

(C>)     $f(C_w) > f(C_v)$     (S>)     $f(S_w) > f(S_v)$

(C<)     $f(C_w) < f(C_v)$     (S<)     $f(S_w) < f(S_v)$

or their weaker variants, for $w > k - w$

(19)     $f(C_w) > f(C_{k-w})$     $(\leftrightarrow f(S_w) > f(S_{k-w}))$

(20)     $f(C_w) < f(C_{k-w})$     $(\leftrightarrow f(S_w) < f(S_{k-w}))$.

Hintikka [16] has argued for a distribution satisfying C> in case the multinomial context is that of random sampling in a denumerably infinite universe which is known to be partitioned by the $k$ Q-predicates. If the context is closed by $C_k$, i.e., if $C_k$ is true, Hintikka speaks of a *Carnapian universe*, in which case he uses also the C-system $f_k$ for assigning probabilities. His proposal is to choose $f(C_W)$ proportional to the probability that (the first) $\alpha$ individuals out of a Carnapian universe turn out to be compatible with $C_W$, i.e., proportional to $f_k(W^\alpha) = \pi(\alpha, w\rho_k)/\pi(\alpha, k\rho_k)$. Together with (16) this leads to the assignment

(H5)     $f(C_w) = \pi(\alpha, w\rho_k) \Big/ \sum_{v=0}^{k} \binom{k}{v} \pi(\alpha, v\rho_k)$.

Note that $f(C_0) = 0$ if $\alpha > 0$, that $f(C_w) = 2^{-k}$ (i.e., CI-B holds) if $\alpha = 0$ and that $f(C_k) = 1$ and $f(C_w) = 0$ for $w < k$ if $\alpha = \infty$. Moreover, it is easy to check that C> holds generally for $0 < \alpha < \infty$.

The so-called $\alpha$-$\lambda$-continuum $(\lambda_w = w\rho_w)$ is an H-system with prior distribution according to H5. The parameter $\alpha$ functions as an index of caution with respect to the constituents $C_w$ with $w < k$: the higher we choose $\alpha$ the lower the prior probability of these constituents and the higher that of $C_k$.

In our opinion H5 has certainly some plausibility in contexts in which $C>$ is plausible. But the latter need not be so, even for an infinite universe. Unfortunately, there seems no way to generalize H5 to a parametric distribution leaving room for, at least, the qualitative relations mentioned above. Moreover, H5 does not lead to an adequate structural distribution when $k \to \infty$, for then $f(S_\infty) = 1$. In view of Chapter 6, Section 10 it is desirable that a parametric distribution leads to $f(S_w) > 0$ $w = 1, 2, \ldots$ for $k \to \infty$, at least under certain conditions.

There are of course many parametric functions with all these properties. But a proposal needs to have some plausibility, such as H5 has. We have presented a parametric distribution [25] leaving room for all mentioned qualitative relations, leading to (two) limit distributions with $f(S_w) > 0$ and, finally, based on an application of GC-systems.

Here we shall only summarize the main aspects of that distribution. Consider the universe of $Q$-predicates, $U_k^Q$. It is, in a particular multinomial context, partitioned by the two properties $E(Q_i): q_i > 0$ and $N(Q_i): q_i = 0$ such that $C_W \leftrightarrow \underset{Q_i \in W}{\&} E(Q_i) \,\&\, \underset{Q_j \in K - W}{\&} N(Q_j)$. In the multinomial context of random sampling in an infinite universe, $E(Q_i)$ comes to: $Q_i$ is exemplified in the universe, and $N(Q_i)$ to: $Q_i$ is not exemplified in the universe.

Now suppose, in the line of Carnap's intended application, that we agree that a GC-system is acceptable as absolute prediction pattern in a context of random sampling without replacement in a finite universe. Suppose further that we are going to draw the elements of $U_k^Q$ one by one, of course, with possible outcomes $E$ and $N$. Then we would have to choose parameters $\gamma_e$, with $0 < \gamma_e < 1$, and $\beta$, with $\beta > 0$ or $-\beta > (k - 1)/\min(\gamma_e, 1 - \gamma_e)$, such that, in Carnap's formulation (1):

(21)    $f(E(Q_{n+1})/e(U_n^Q)) = (n_e + \gamma_e\beta)/(n + \beta), \quad n < k.$

By the product rule we obtain from (21)

(EN)    $f(C_w) = \pi(w, \gamma_e\beta) \cdot \pi(k - w, (1 - \gamma_e)\beta)/\pi(k, \beta), \quad 0 \le w \le k.$

In a multinomial context we are not, however, sampling in $U_k^Q$ but this need not make any difference to our prior expectation (or belief) in $C_w$. Hence, EN is a plausible constitutional distribution.

For $\beta = \infty$, EN turns into

(22)     $f(C_w) = (\gamma_e)^w (1 - \gamma_e)^{k-w}$

and $f(S_w)$ becomes then, on the basis of (14), a so-called binomial distribution.

The symmetric EN-distributions are those for which $\gamma_e = \frac{1}{2}$. If, in addition, $\beta = \infty$ we get CI-B, if $\beta = 2$ we get SI-B (the distribution suggested by Carnap), and for $\beta = -2k$ we get

(23)     $f(C_w) = \binom{k}{w} \Big/ \binom{2k}{k}$.

In the first case $f(S_w)$ is proportional to the number of $w$-constituents. In the second case $f(C_w)$ is inversely proportional to that number. Finally, in the third case $f(C_w)$ is proportional to that number.

As to the non-symmetric distributions we consider only the case $\gamma_e > \frac{1}{2}$ (and $\beta > 0$) for all relations to be given change in the opposite for $\gamma_e < \frac{1}{2}$. Now, if $\gamma_e > \frac{1}{2}$ and $\beta > 0$, then (19) holds generally. Moreover we have

(24)     $C>$ if and only if $\beta(2\gamma_e - 1) > k - 1$

(25)     $S>$ if (not only if) $1/\gamma_e < \beta < 1/(1 - \gamma_e)$.

It is easy to check that $C>$ and $S>$ can both be satisfied.

In Chapter 5, Section 3.3 we showed that a GC-system can be interpreted as the generalization of an urn-model described by Polya. Polya [30] has given a number of limit distributions for that model. We list only the discrete limit distributions, based on EN and (14).

(26)     If $k \to \infty$, $\beta \to \infty$, $\gamma_e \to 0$ such that $k\gamma_e \to \mu > 0$ and $k/\beta \to 0$, then $f(S_w) = e^{-\mu}\mu^w/w!$.

This distribution is called the Poisson distribution

(27)     If $k \to \infty$, $\beta \to \infty$, $\gamma_e \to 0$ such that $k\gamma_e \to \mu > 0$ and $k/\beta \to 1/r > 0$, then
$$f(S_w) = \binom{r\mu + w - 1}{w} \left(\frac{r}{1+r}\right)^{r\mu} \left(\frac{1}{1+r}\right)^w.$$

This distribution is a special case of the so-called negative binomial distribution (see Feller [10], p. 165).

## 6. THE HYPERGEOMETRIC CONTEXT

Let there be a finite universe with $N$ elements and which is known to be partitioned by the finite set $K$ of $Q$-predicates. There are therefore, unknown, numbers $N_i$, $i = 1, 2, \ldots, k$, $\sum_i N_i = N$, such that $N_i$ is the number of $Q_i$-elements in the universe. Our experiments are successive random samplings *without replacement* in that universe. These experiments are governed by the so-called (multiple) hypergeometric distribution, with special values

(28)    $P(Q_i/e_n) = (N_i - n_i)/(N - n), \quad n < N, n_i \leq N_i.$

This objective pattern is an improper negative GC-system with parameters $-\lambda = \eta = N$ and $\gamma_i = N_i/N$. It is improper because it does not satisfy the parameter condition $(-)(\eta \geq (N - 1)/\gamma_i$ for all $i)$ and, consequently, it leads to 'negative probabilities' if it is considered as a pattern with respect to $K, K^2, \ldots, K^N$. This is of course not objectionable for an objective pattern, for the events with negative probability cannot occur.

Apart from the improper character, $P$ satisfies the GC-principles POI, PRR and GPL but not necessarily PIP.

More or less analoguous to the multinomial context we take as elementary theories $C_W(N)(W \subset K)$, stating that $W = \{Q_i \in K/N_i > 0\}$. By consequence, $e_N$ is compatible with $C_W(N)$ if and only if $e_N \in H_W(N)$, the $C_W(N)$-spaces are $W, W^0, \ldots, W^N$ and the elementary theories are decidable.

The described decidable context will be called a *hypergeometric context*. In Section 1 we have argued that this is in fact the context of application intended by Carnap for C-systems.

Let us agree that we need a PER-system as an expectation pattern, for at the start we do not know of differences between the predicates whereas the underlying process does suggest that the order of outcomes does not give relevant information.

The conditional patterns have to be closed, i.e., in the present context $f_W(H_W(N))$ has to be 1. It is easy to see that we cannot obtain this with a C-system, as a conditional pattern, for in such a system it holds that, even if $e_N \in W^N$ is such that $n_i(e_N) = N$ for some $i$, then $f_W(e_N) > 0$.

This problem cannot arise if we start from an absolute pattern, for in that case the analogue of (13) guarantees that $f_W(e_N) = 0$ for all $e_N \in (W^N - H_W(N))$. Following Carnap we now consider a C-system as an absolute pattern, but let us leave room for the possibility that $\lambda$ is negative (but then satisfying the parameter condition, which comes to: $-\lambda \geq (N - 1) . k$) and that it may not only depend on $k$ but also on $N$. The latter dependency was already argued for in Section 1.

Up to now we did not answer the fundamental question what we do want to learn from experience, i.e., what objective state of affairs do we want to approach. As to the posterior belief function the situation is unproblematic, for any absolute pattern leads to $f(C_M(N)/e_N) = 1$, which seems very welcome. Unfortunately, the situation with regard to the (absolute) special values is rather difficult. To begin with, the objective special value function (28) defines no value for $n = N$ (and $n_i = \mathbf{N}_i$ for all $i$). Consider therefore $e_{N-1}$. There must be one $Q_i$ for which $n_i(e_{N-1}) = \mathbf{N}_i - 1$, whereas for all $Q_j$, $i \neq j$, $n_j(e_{N-1}) = \mathbf{N}_j$. From (28) we obtain that $P(Q_i/e_{N-1}) = 1$ and $P(Q_j/e_{N-1}) = 0$ for all $j \neq i$. The corresponding values in a C-system are $(\mathbf{N}_i - 1 + \lambda/k)/(N - 1 + \lambda)$ and $(\mathbf{N}_j + \lambda/k)/(N - 1 + \lambda)$, respectively. It is easy to see that there is no way to choose $\lambda$ such that the latter special values approach 1 and 0, respectively, irrespective of the $\mathbf{N}_i$, $\mathbf{N}_j$, $j \neq i$.

One might suggest that $f(Q_i/e_n)$ needs to approach $P(Q_i) = \mathbf{N}_i/N$ and this is what a (positive) C-system actually does. But the prize is that we are no longer taking seriously the attempt to assign rational probabilities to future events. Our conclusion is that a C-system is adequate in approaching the *original* state of affairs but inadequate as an expectation pattern. As far as we know, Carnap has nowhere disentangled these two objectives which turn out to be incompatible in the present type of context. Note that this divergence of objectives does not occur in a multinomial context for the simple reason that in the latter context the system does not change.

We conclude this section with some tentative suggestions for a rational expectation pattern w.r.t. a hypergeometric context. The argument, given in Section 1, in favour of a dependency of $\lambda$ on $N$, supports the requirement that the special values depend on $N$. Moreover, the objective pattern has the property of negative instantial relevance:

(29)     $P(Q_i/e_n Q_i) < P(Q_i/e_n), \quad n < N, n_i < \mathbf{N}_i - 1.$

In our opinion it is desirable that the special values of our expectation pattern also have this property for $n$ near to $N$ and $n_i$ small. On the other hand, for small $n$ we are inclined to require, generally, the opposite: positive instantial relevance.

# CHAPTER 8

# SOME PROBLEMS AND RELATED TOPICS

In the first three sections of this chapter we formulate some technical problems which we have tried to solve in our research, but without success. All these problems may or may not have solutions of philosophical importance.

In the other three sections we pay some attention to topics which are strongly related to the present study. Sections 4 and 5 are both related to the original point of departure of Carnap's explication program: the evaluation of theories in the light of (new) evidence, i.e., confirmation and falsification. In the last section we shall make some remarks about proposals of Hintikka and Hilpinen for rules of acceptance in UC-systems.

## 1. PER-SYSTEMS

In Section 4 of Chapter 7 we have defined PER-systems in terms of restrictions to a formally rational expectation pattern w.r.t. to a multinomial context. But, it is of course also possible to give that definition in purely mathematical terms: a PER-system is a regular consistent probability terms: a PER-system is a regular consistent probability pattern w.r.t. $K, K^2, \ldots$ such that

$$(1) \qquad p(e_n) = f(n_1, n_2, \ldots, n_k).$$

((1) is supposed to imply that $p(e_n)$ is not only invariant for order-permutations but also for predicate-permutations.) A PER-system is said to be *closed* if $p(H_k) = 1$ and *open* if $p(H_k) < 1$. Note that $p(H_k) > 0$ is implied by the requirement of regularity. In an open PER-system the conditional patterns are defined by

$$(2) \qquad p_W(e_n) = p(e_n/H_W) = {}_{df} p(e_n \cap H_W)/p(H_W), \quad e_n \in W^n,$$

provided that $p(H_W) \neq 0$. It is clear that these conditional patterns are closed PER-systems (w.r.t. $W$, $W^2$, ...):

(3)     $p_W(H_W) = 1$.

In the original introduction of H-systems in Section 3 of Chapter 6 we constructed an H-system on the basis of a probability function $q$ w.r.t. the $H_W$'s, conditional patterns $q_W$ w.r.t. $W$, $W^2$, ... and definition (18). In that set-up (assuming $q(H_k) < 1$) we also get, of course, an open PER-system as soon as we require that

(4)     $q(H_W) = q(H_V)$,  if $v = w$

(i.e., H3) and that the conditional patterns need to be PER-systems. If the latter are closed we have

(5)     $p(H_W) = q(H_W)$

(6)     $p_W(e_n) = q_W(e_n)$

in which case we talk about the standard representation of that PER-system. If the conditional patterns are open, then (5) and (6) no longer hold, but it follows directly from (2) and (3) that there is a standard representation. In what follows we shall assume the standard representation of open PER-systems. Note that a closed PER-system cannot be formulated in a non-standard way for then $1 = p(H_k) = q(H_k)q_k(H_k)$ (based on (18) of Section 3, Chapter 6) leads to 'probability values' exceeding 1.

C-systems are of course closed PER-systems and H-systems (and therefore UC-systems), with $p(H_k) < 1$, are open PER-systems with C-systems as (closed) conditional patterns. The class of C-systems determines, in fact, all possible conditional patterns for H-systems. In the same way we have that the class of closed PER-systems not being C-systems determines the class of open PER-systems not being H-systems.

This brings us to the, at least mathematically, intriguing question of a general parametric characterization of closed PER-systems with C-systems as special cases. Unfortunately, we did not find such a characterization. We have also payed attention to the more restricted question of a parametric characterization of closed PER-systems satisfying

(7)     $p(Q_i/e_n) = (n_i + \rho(e_n))/(n + k\rho(e_n))$

in which $\rho(e_n)$ is supposed to be a function depending at most on all $n_i$. The restriction to (7) was made in view of the fact that (7) holds for

C-systems in the trivial sense that $\rho(e_n)$ is a constant. Again, we did not obtain results worth being listed here.

## 2. ON WEAKENING WPERR

In Section 8 of Chapter 6 we formulated the problem of a parametric characterization of H-systems in terms of finite principles. In the terminology of PER-systems, UC-systems are, in their NH-formulation, PER-systems satisfying WPERR:

(8)     $p(Q_i/e_n) = f_m(n, n_i)$.

At the same time they could be formulated as Special H-systems. The question is whether WPERR can be weakened in such a way that we obtain the whole class of H-systems. We have tried this by looking to the NH-principles, which were in conjunction, equivalent to WPERR. It is easy to check that NH2.1 $(p(Q_i/e_n\overline{M}) = 1/(k - m), Q_i \notin M)$ is already contained in the requirement that we want to have a PER-system.

As for NH1 and NH2.2 we consider the relevant values for H-systems. First we define

(9)     $a_w(e_n) = _{\text{df}} \left( \dfrac{k - m}{w - m} \right) p(H_w) \dfrac{\prod_i \pi(n_i, \rho_w)}{\pi(n, w\rho_w)}, \quad m \le w \le k$

(10)    $R(e_n) = _{\text{df}} \displaystyle\sum_{w=m}^{k} a_w(e_n)\rho_w/(n + w\rho_w)$

(11)    $S(e_n) = _{\text{df}} \displaystyle\sum_{w=m}^{k} a_w(e_n)/(n + w\rho_w)$.

Now it is easy to check that in an H-system holds

(12)    $p(M/e_n) = (nS(e_n) + mR(e_n))/p(e_n)$

in which $p(e_n) = \displaystyle\sum_{w=m}^{k} a_w(e_n)$, and

(13)    $p(Q_i/e_nM) = (n_iS(e_n) + R(e_n))/(nS(e_n) + mR(e_n)), \quad Q_i \in M$.

In a UC-system $R(e_n)/S(e_n)$ is simply the constant $\rho$. From (12) and (13) it follows that $p(Q_i/e_n)$ can be written as

(14)    $p(Q_i/e_n) = n_iA(e_n) + B(e_n), \quad n_i > 0$.

Conversely, (14) may now be seen as a weak version of **WPERR** if $A(e_n)$ and $B(e_n)$ are limited to functions depending only on $n_1, \ldots, n_k$. Analogous to the NH-principles (14) can be disentangled into

(15)      $p(M/e_n) = nA(e_n) + mB(e_n)$

(16)      $p(Q_i/e_n M) = (n_i A(e_n) + B(e_n))/(nA(e_n) + mB(e_n))$.

However, (14) (or (15) and (16)) seems too weak to characterize the subclass of H-systems within the class of (open) PER-systems and we did not succeed in finding the necessary restriction to (14) (or to (15) and (16)).

### 3. *UC*-SYSTEMS AND $k \to \infty$

We define *the* system $^{*}UC_k^{*}$ as the UC-system in which $\rho = 1$ and $p(Z_w) = \binom{k}{w} p(H_w) = 1/k$ (i.e., the distribution suggested by Carnap). Using the reformulation of H-systems in Section 10 of Chapter 6, we have, on the basis of (82), (84) and (85) in $^{*}UC_k^{*}$:

(17)      $\mathbf{p}(Z_m/\mathbf{e}_n) = 1 \Big/ \displaystyle\sum_{w=m}^{k} \binom{w}{m} \frac{\pi(n, m)}{\pi(n, w)}$

$= 1 \Big/ \displaystyle\sum_{w=m}^{k} \frac{w!}{m!(w-m)!} \cdot \frac{(m+n-1)!}{(w+n-1)!} \cdot \frac{(w-1)!}{(m-1)!}$.

Note that $\mathbf{p}(Z_m/\mathbf{e}_n)$ corresponds to $p(Z_m/e_n)$ in the original formulation (use (40) and (74)).

If $k \to \infty$ then $\mathbf{p}(Z_w) \to 0$, and therefore $^{*}UC_\infty^{*}$ seems to collapse completely. The following observation may however be of fundamental importance. It is easy to show, on the basis of (17), that, for $m(\mathbf{e}_3) = 1$,

(18)      $\mathbf{p}(Z_1/\mathbf{e}_3) = (2 + k)/(3k)$

and, therefore, that

(19)      $\mathbf{p}(Z_1/\mathbf{e}_3) \to 1/3$  if $k \to \infty$.

Unfortunately, we did not find a general expression, in $n$ and $m$, for the positive limits of $\mathbf{p}(Z_m/\mathbf{e}_n)$ for $k \to \infty$. But we do not see any reason why (19) should be completely accidental. Therefore, our hypothesis is that $^{*}UC_\infty^{*}$ does not collapse rigorously, although it is not (and cannot be) a regular probability pattern. If this hypothesis is true, this might be seen as an argument in favour of the distribution SI-B compared with CI-B (see Section 5 of Chapter 7) for it is easy to show that a UC-system based on the *latter* collapses completely if $k \to \infty$.

## 4. CONFIRMATION THEORY

In Chapter 2 we discussed the general philosophical claims of inductive probability theory. There we did not pay attention to the particular claim to provide a contribution to confirmation theory. Confirmation theory is still in its infancy: it can be seen as an explication-program, within philosophy of science, based on the intuitive statement that theories can be confirmed or disconfirmed by (new) evidence (without being verified or falsified).

Carnap in his earlier work [2, 3] still thought that the intuitive notions of rational degree of belief (and expectation) and degree of confirmation were the same. Due to the criticism of Popper and others, Carnap declared in the preface to the second edition of his *Logical Foundations of Probability* [2], published in 1962, to have conflated two different intuitions. Batens [1] has given a lucid account of the different intuitions that played a role in the controversy between Popper and Carnap.

According to Batens there are at least three different, but related, intuitive notions in this connection. In our terminology, the first is that of (rational) degree of expectation and this notion has to be explicated as a probability value. The other two are degree of confirmation and degree of corroboration. Both are directed to confirmation and have to be explicated in terms of *changes* in probability values. Poppers notion of degree of corroboration includes not only aspects of confirmation but also aspects of the content of theories: it is at least a mixture of the intuitive notions of degree of confirmation and degree of content. We agree with Batens that the notion of a degree of confirmation in a strict or pure sense is worth being explicated.

In our opinion, Batens' own proposal for a pure confirmation function is very attractive. Within (the symbolisim of) a formally rational expectation pattern w.r.t. a paradigmatic context (Chapter 4) the proposal comes to about

$$(20) \qquad K(T, E_n) =_{\text{df}} \frac{f_T(E_n)}{f_T(E_n) + f_{D-T}(E_n)}$$

and it is easy to prove that this definition is equivalent to

$$(21) \qquad K(T, E_n) =_{\text{df}} \cfrac{1}{1 + \cfrac{f(T/E_n)^{-1} - 1}{f(T)^{-1} - 1}}.$$

The following properties of this degree of confirmation of $T$ by $E_n$ are easy to prove.

(22) $\quad 0 \leqslant K(T, E_n) \leq 1$

(23) $\quad K(T, E_n) + K(D - T, E_n) = 1$

(24) $\quad$ if $f(T/E_n) = 0$ then $K(T, E_n) = 0$ $\qquad$ (falsification)

(25) $\quad$ if $0 < f(T/E_n) < f(T)$ then $0 < K(T, E_n) < \frac{1}{2}$ $\quad$ (disconfirma-
tion)

(26) $\quad$ if $f(T, E_n) = f(T)$ then $K(T, E_n) = \frac{1}{2}$ $\qquad$ (neutral
confirmation)

(27) $\quad$ if $f(T) < f(T/E_n) < 1$ then $\frac{1}{2} < K(T, E_n) < 1$ $\quad$ (confirmation)

(28) $\quad$ if $f(T/E_n) = 1$ then $K(T, E_n) = 1$. $\qquad$ (verification)

The condition for falsification is satisfied as soon as $E_n$ is incompatible with $T$. The condition for verification is satisfied if and only if $T$ contains all elementary theories compatible with $E_n$. Finally, the condition for neutral confirmation is satisfied for $E_n = V_n$ or, in other words, before we start the experiments.

We conclude this section with a natural extension of Batens' confirmation function.

$$(29) \qquad K(T, E_n(k); E_k) =_{\text{df}} \frac{f_T(E_n(k)/E_k)}{f_T(E_n(k)/E_k) + f_{D-T}(E_n(k)/E_k)}.$$

A formulation similar to (21) is obtained by replacing in that definition $f(T/E_n)$ by $f(T/E_k E_n(k))$ and $f(T)$ by $f(T/E_k)$. It is also easy to check that $K(T, E_n(k); E_k)$ has the same properties ((22)–(28)) as $K(T, E_n)$ if the mentioned replacements are made in the conditions. Note that the properties of universal confirmation defined in Section 4 of the preceding chapter (PCf-B and PCf-Bs) are in accordance with the present terminology, e.g., PCf-B amounts to $\frac{1}{2} < K(C_M, M; e_n) < 1$, i.e., given $e_n$, $C_M$ is confirmed by $M$.

$K(T, E_n(k); E_k)$ provides a measure for the role of new evidence in the confirmation of a theory. Its structure in relation to the absolute degrees of confirmation can be clarified in terms of the ratio of the latter degrees:

(30)     $K_r(T, E_n(k); E_k) =_{df} K(T, E_k E_n(k))/K(T, E_k)$

for (29) can now be rewritten as

(31)     $K(T, E_n(k); E_k) = \dfrac{K_r(T, E_n(k); E_k)}{K_r(T, E_n(k); E_k) + K_r(D - T, E_n(k); E_k)}$

We have proposed [22] an explicative definition of Poppers notion of degree of corroboration on the basis of Batens' confirmation function. On the same basis we have shown [24] that Poppers so-called paradox of ideal evidence ([31], Appendix *ix) is not a paradox at all.

## 5. FALSIFICATION

Lakatos [27] has given a profound analysis of falsification. He distinguishes a naive and a sophisticated conception of falsification and argues that Popper presupposes at some places the naive conception but that he is, at other places, well aware of the sophisticated one. Although Lakatos intended the analysis certainly not for such simple scientific situations, as a multinomial context in fact is, it is (even in general) not without value to know whether or not such an analysis works in simple cases.

According to Lakatos ([27], p. 116) scientific theory $T$ is falsified in the sophisticated sense 'if and only if another theory $T'$ has been proposed with the following characteristics:

(a)     $T'$ has excess empirical content over $T$: that is, it predicts *novel* facts, that is, facts improbable in the light of, or even forbidden by. $T$;

(b)     $T'$ explains the previous success of $T$, that is, all the unrefuted content of $T$ is included (within the limits of observational error) in the content of $T'$; and

(c)     some of the excess content of $T'$ is corroborated'.

Suppose now that, in a multinomial context, the first $n$ experiments result in evidence $e_n$. Let the next experiment result in some $Q_i$ not in $M(e_n) = M$ and let $M' = M \cup \{Q_i\}$. If we now substitute in the definition of Lakatos (the constituent) $C_M$ for $T$ and $C_{M'}$ for $T'$ then it is easy to check that $C_{M'}$ has, in any conceivable reading, all characteristics required by the definition, i.e., $C_M$ has been falsified (by the $(n + 1)$-th experiment) in the sophisticated sense.

Of course in this situation $C_M$ is also falsified in the naive sense: the evidence $e_n Q_i$ is incompatible with $C_M$. This is, however, not surprising in the light of clauses (a) and (c) of the definition of sophisticated falsification, for they give room for such new evidence.

## 6. RULES OF ACCEPTANCE IN UC-SYSTEMS

In Section 3 of Chapter 2 we announced already a rule of acceptance for a specific context proposed by Hintikka and Hilpinen. This rule was formulated for $x$–$\lambda$-systems being UC-systems [18]. Hilpinen ([15], Chapter 5.4) has weakened that rule. In our terminology, both rules are prepared for the multinomial context of random sampling (without replacement) in a partitioned universe with infinitely many individuals. In this specific context it is plausible to call a theory (i.e., a set of constituents) also a (universal) generalization and an hypotheses about the outcome of the next experiment a singular hypothesis. We shall adopt this terminology in what follows.

The restriction of Hintikka and Hilpinen to $\alpha$–$\lambda$-systems will not be made here, i.e., both rules will be formulated for any UC-system. Moreover, the verbal reference to the specific context sketched above, is easily shown to be inessential, i.e., everything could be formulated purely in terms of UC-systems.

Consider for some fixed $\epsilon$, $0 < \epsilon < \frac{1}{2}$, the condition

(32)      $f(C_M / e_n) > 1 - \epsilon.$

Using (40) of Chapter 6, this is equivalent to

$$(33) \qquad \sum_{w=m+1}^{k} \binom{k-m}{w-m} \frac{f(C_w)\pi(n, m\rho)}{f(C_M)\pi(n, w\rho)} < \epsilon/(1 - \epsilon) < 1.$$

Define now $n_m$ as the largest integer $n$ for which the converse of (33) holds. This integer exists, for $\pi(n, m\rho)/\pi(n, w\rho) \to 0$ if $n \to \infty$ and $w > m$ (see the proof of T3 of Chapter 5, Section 4.2). Hence, $n_m$ depends on $\epsilon$, $m$, $\rho$

and the constitutional distribution and (33) becomes equivalent to $n > n_m$.

The weak acceptance rule, for an arbitrary generalization $G$, i.e., a disjunction of a number of $C_W$, now reads

(WAG)  Accept $G$, on the basis of $e_n$, if and only if
  (1)  $f(G|e_n) > 1 - \epsilon$         (2)  $n > n_m$.

It is easy to show that WAG is equivalent to

(WAG′)  Accept $G$ on the basis of $e_n$ if and only if
    $C_M$ may be accepted according to WAG and
    $C_M$ is a member (a disjunct) of $G$.

The original (stronger, but more cautious) rule of acceptance (SAG) is obtained by replacing WAG(2) by

(SAG)(2)  $n > \max_m (n_m)$

and SAG′ is defined similar to WAG′. SAG is called stronger than WAG because SAG strengthens the condition on $n$ in such a way that it becomes independent of $m$. It is more cautious because SAG-acceptability implies WAG-acceptability and the converse does not hold.

Both rules have a natural extension to singular hypotheses. Let $G_W$ be defined as the generalization containing precisely the constituents $G_V$ for which $V \subset W$.

(AS)    Accept, on the basis of $e_n$, the hypothesis that the next individual to be drawn belongs to $W$, if and only if $G_W$ may be accepted (according to WAG or SAG).

This rule is, on the basis of WAG′ and SAG′, equivalent to

(AS′)   Accept, on the basis of $e_n$, the hypothesis that the next individual to be drawn belongs to $W$, if and only if $W \supset M$ and $C_M$ may be accepted (according to WAG or SAG).

It is not difficult to prove that WAG (or SAG) and AS satisfy the two requirements mentioned at the beginning of Section 3 of Chapter 2: the set of acceptable generalizations and singular hypotheses is consistent and deductively closed. By consequence, the lottery-paradox cannot arise.

In our opinion Hilpinen is right in his implicit suggestion that WAG is preferable to SAG despite the fact that the former is less cautious than the latter. The present general set-up in terms of UC-systems suggests even a further refinement of WAG. Thusfar, $\epsilon$ was a fixed number. Let there now be numbers $\epsilon_w (0 < \epsilon_w < \frac{1}{2}, w = 1, 2, \ldots, k)$. Define $n_m$ as before, but now with replacement of $\epsilon$ by $\epsilon_m$ in (33). Finally, define $\text{WAG}_m$ as WAG but with replacement of $\epsilon$ by $\epsilon_m$. It is easy to show that the conditions of consistency and deductive closedness remain satisfied.

It seems natural that $\epsilon_w$ is now functionally related to the constitutional distribution (and perhaps $\rho$) in general and to $f(C_w)$ in particular. However, we did not find a satisfactory proposal for such a function.

CHAPTER 9

## CONCLUDING REMARKS

In order to contribute to an adequate understanding of the program of explicating cognitive rationality we shall consider three questions. The first is, whether there are prospects for the construction of rational expectation patterns in other than multinomial contexts. The second is, whether and in what sense, the program may be said to be logical. The third is, whether the program can lead to a substantial contribution to the philosophy of science.

The first question presupposes our conclusion that the application of C- and UC-systems to closed and open multinomial contexts, respectively (Chapter 7), are fundamentally adequate. In our opinion these applications have to be considered in at least two respects as paradigmatic examples for further research. The general character of these examples shows that the set-theoretical approach is superior to Carnap's logico-linguistic approach. Moreover, they suggest we should select, for further study, contexts in which there is a well-defined underlying probability process, about which only the structure is known beforehand. The hypergeometric context (Chapter 7, Section 6) is a case in point. Other examples, which seem more easy to handle, are contexts based on Markov-processes and on changing multinomial processes. A concrete example of the latter is a coloured ball on which the partitioning changes according to some probability process, e.g., a process describable in terms of a GC- or even a GH-system. In case of a GC-system this example is almost equivalent to the generalizaton of Polya's urn-model (Chapter 5, Section 4.3).

In our opinion, research directed to these contexts has a good chance of being successful. But, unfortunately, we do not see in what way these examples can be more than curiosities as far as philosophy is concerned.

Our second question is of a classificatory nature. Carnap's abandonment of inductive rules of inference (see Chapter 2, Section 3) made the term 'inductive logic' already problematic, but it continued to have some plausibility in his logico-linguistic approach. In the set-theoretical approach even this is no longer the case. In Chapter 2 we already expressed our preference to speak of (pure and applied) inductive probability theory for that reason.

There is, however, one analogy between inductive probability theory and (deductive) logic which deserves to be mentioned. Syntactic approaches in logic are based on the formal structure of sentences and arguments. The systems that have been studied here are all essentially based on structural aspects of theories and evidence. Perhaps there is here a possibility to give a definition of pure inductive probability theory which is as rigorous as the distinction that could be made (Chapter 2, Section 3) between objective and inductive applications of probability theory.

Of course the term inductive logic continues to have some plausibility if appropriate rules of acceptance are added to expectation patterns. But because of the fact that they lack, necessarily, the truth-preserving character and because their most general formulation, in a particular type of context, is again set-theoretical, it may be better to consider their study as a special branch of (inductive) probability theory. An additional reason for doing so, is that such rules seem to be highly specific for a particular type of context as is suggested by the rules studied in Chapter 8, Section 6.

With respect to the last question, the significance for the philosophy of science, we shall be very brief. The debate about the desirability of a (sophisticated) theory of confirmation has a long history. Notably Popper has stressed the superfluity of such a theory. He and his followers have defended that the process of science has to be explicated in terms of rejection of tentatively proposed theories. For the evaluation of theories, qualitative, including comparative, notions of corroboration are sufficient.

One might object to this view that a quantitative degree of confirmation may still have some value in what Kuhn has called normal science. But, even in normal science, most theories and relevant experiments do not look like multinomial contexts or the above mentioned contexts for which the construction of expectation patterns was said to have a good chance.

To summarize our discussion of the first and the last question, our opinions about the philosophical relevance of extending the explication program are rather pessimistic, despite our optimism about the possibility of extension. However, this does not exclude that inductive probability theory may be useful in other areas, such as statistics and learning theory. But this has to be judged by others.

# REFERENCES

[1] Batens, Diderik, 'Some Proposals for the Solution of the Carnap-Popper Discussion on "Inductive Logic" ', *Studia Philosophica Gandensia* (1968), pp. 5–25.

[2] Carnap, Rudolf, *Logical Foundations of Probability*, University of Chicago Press, Chicago, 1950, 1963².

[3] Carnap, Rudolf, *The Continuum of Inductive Methods*, University of Chicago Press, Chicago, 1952.

[4] Carnap, Rudolf, 'The Aim of Inductive Logic', in *Logic, Methodology and Philosophy of Science* ed. by E. Nagel, P. Suppes and A. Tarski), Stanford University Press, Stanford, 1962, pp. 303–318.

[5] Carnap, Rudolf, 'The Concept of Constituent-Structure', in *The Problem of Inductive Logic* (ed. by I. Lakatos), North-Holland Publishing Co., Amsterdam, 1968, pp. 218–220.

[6] Carnap, Rudolf, 'Inductive Logic and Rational Decisions', in *Studies in Inductive Logic and Probability Vol. I* ed. by R. Carnap and R. C. Jeffrey), University of California Press, Berkeley, 1971, pp. 5–31.

[7] Carnap, Rudolf, 'A Basic System of Inductive Logic, Part I', in *Studies in Inductive Logic and Probability Vol. I* (ed. by R. Carnap and R. C. Jeffrey), University of California Press, Berkeley, 1971, pp. 33–165.

[8] Carnap, Rudolf and Stegmüller, Wolfgang, *Induktive Logik und Wahrscheinlichkeit*, Springer-Verlag, Wien, 1959.

[9] Essler, Wilhelm K., 'Hintikka versus Carnap' in [17], pp. 365–369.

[10] Feller, William, *An Introduction to Probability Theory and Its Applications Vol. I*, John Wiley & Sons, New York, 1950, 1968³.

[11] Feller, William, *An Introduction to Probability Theory and Its Applications Vol. II*, John Wiley & Sons, New York, 1964, 1968².

[12] Friedman, B., 'A Simple Urn Model', *Communications on Pure and Applied Mathematics* 2 (1949), 59–70.

[13] Hempel, Carl G., *Fundamentals of Concept Formation in Empirical Science*, University of Chicago Press, Chicago, 1952.

[14] Hempel, Carl G., 'Inductive Inconsistencies', *Synthese* 12 (1960), 439–469, reprinted in *Aspects of Scientific Explanation*, The Free Press, New York, 1965, pp. 53–79.

[15] Hilpinen, Risto, *Rules of Acceptance and Inductive Logic* (*Acta Philosophica Fennica* 22), North-Holland Publishing Co., Amsterdam, 1968.

[16] Hintikka, Jaakko, 'A Two-dimensional Continuum of Inductive Methods', in *Aspects of Inductive Logic* ed. by J. Hintikka and P. Suppes), North-Holland Publishing Co., Amsterdam, 1966, pp. 113–132.

[17] Hintikka, Jaakko (ed.), *Rudolf Carnap, Logical Empiricist*, D. Reidel Publishing Co., Dordrecht-Holland, 1975.

[18] Hintikka, Jaakko and Hilpinen, Risto, 'Knowledge, Acceptance and Inductive Logic', in *Aspects of Inductive Logic* (ed. by J. Hintikka and P. Suppes), North-Holland Publishing Co., Amsterdam, 1966, pp. 96–112.

[19] Hintikka, Jaakko and Niiniluoto, Ilkka, 'An Axiomatic Foundation for the Logic of Inductive Generalization', in [32], pp. 57–81.

[20] Howson, C., 'Must the Logical Probability of Laws be Zero?' *Brit. J. Phil. Sci.* **24** (1973), pp. 153–163.

[21] Kemeny, John G., 'Carnap's Theory of Probability and Induction', in *The Philosophy of Rudolf Carnap* (ed. by P. A. Schilpp), Open Court, LaSalle, 1963, pp. 711–738.

[22] Kuipers, Theo A. F., 'A Note on Confirmation', *Studia Philosophica Gandensia* **10** (1972), pp. 76–77.

[23] Kuipers, Theo A. F., 'A Generalization of Carnap's Inductive Logic', *Synthese* **25** (1973) pp. 334–336 (reprinted in [17]).

[24] Kuipers, Theo A. F., 'Inductive Probability and the Paradox of Ideal Evidence', *Philosophica* **17.1** (1976), pp. 197–205.

[25] Kuipers, Theo A. F., 'A Two-dimensional Continuum of *a priori* Probability Distributions on Constituents', in [32], pp. 82–92.

[26] Kuipers, Theo, A. F., 'On the Generalization of the Continuum of Inductive Methods to Universal Hypotheses', to appear in *Synthese* **37.1** (1978).

[27] Lakatos, Imre, 'Falsification and the Methodology of Scientific Research Programmes', in *Criticism and the Growth of Knowledge* (ed. by I. Lakatos and A. Musgrave), Cambridge University Press, Cambridge, 1970, pp. 91–196.

[28] Mooij, J. J. A., *Aspecten van redelijkheid*, Wolters-Noordhoff, Groningen-Holland, 1971.

[29] Pietarinen, Juhani, *Lawlikeness, Analogy and Inductive Logic* (Acta Philosophica Fennica 26), North-Holland Publishing Co., Amsterdam, 1972.

[30] Polya, G., 'Sur quelques points de la théorie des probabilités', *Annales de l'Institut Henri Poincaré* **1** (1931), pp. 117–161.

[31] Popper, Karl R., *The Logic of Scientific Discovery*, Hutchinson & Co, London, 1959.

[32] Przełecki, Marian, Szaniawski, Klemens and Wójcicki, Ryszard (eds), *Formal Methods in the Methodology of Empirical Sciences*, Ossolineum, Warszawa, 1976 (also Synthese Library 103).

[33] Shimony, Abner, 'Coherence and the Axioms of Confirmation', *Journal of Symbolic Logic* **20** (1955), pp. 1–28.

[34] Stegmüller. Wolfgang, *Carnap II: Normative Theorie des induktiven Räsonierens*, Part C, Volume IV, Probleme und Resultate der Wissenschaftstheorie und Analytischen Philosophie. Springer-Verlag, Berlin, 1973.

# INDEX OF NAMES

# INDEX OF SUBJECTS

*Recurring symbols* (w.r.t. $K$, $K^2$, . . .)

$e_n \in K^n$, $E_n \subset K^n$

$n_i(e_n)$: number of $Q_i$'s in $e_n$

$\ddot{e}_n = \{e_n'/\forall_i n_i(e_n') = n_i(e_n)\}$

$W \subset K$: $|W| = w$

$M(e_n) = \{Q_i/n_i(e_n) > 0\}$

$m(e_n) = |M(e_n)|$

$e_m{:}e_m \in K^m$ and
$\quad |M(e_m)| = m$, $\quad m = 1, 2, \ldots, k$

$H_W(n) = \{e_n/M(e_n) = W\}$

$H_W = \bigcup\limits_{n=w}^{\infty} H_W(n)WWW \ldots$

$Z_w = \bigcup\limits_{|W|=w} H_W$

$\pi(n, x) = x(x + 1) \ldots (x + n - 1)$
$\quad n = 1, 2, \ldots$

$\pi(0, x) = 1$

145

# SYNTHESE LIBRARY

Monographs on Epistemology, Logic, Methodology,
Philosophy of Science, Sociology of Science and of Knowledge, and on the
Mathematical Methods of Social and Behavioral Sciences

*Managing Editor:*
JAAKKO HINTIKKA (Academy of Finland and Stanford University)

*Editors:*

ROBERT S. COHEN (Boston University)
DONALD DAVIDSON (University of Chicago)
GABRIËL NUCHELMANS (University of Leyden)
WESLEY C. SALMON (University of Arizona)

1. J. M. Bocheński, *A Precis of Mathematical Logic.* 1959, X + 100 pp.
2. P. L. Guiraud, *Problèmes et méthodes de la statistique linguistique.* 1960, VI + 146 pp.
3. Hans Freudenthal (ed.), *The Concept and the Role of the Model in Mathematics and Natural and Social Sciences, Proceedings of a Colloquium held at Utrecht, The Netherlands, January 1960.* 1961, VI + 194 pp.
4. Evert W. Beth, *Formal Methods. An Introduction to Symbolic Logic and the Study of Effective Operations in Arithmetic and Logic.* 1962, XIV + 170 pp.
5. B. H. Kazemier and D. Vuysje (eds.), *Logic and Language. Studies Dedicated to Professor Rudolf Carnap on the Occasion of His Seventieth Birthday.* 1962, VI + 256 pp.
6. Marx W. Wartofsky (ed.), *Proceedings of the Boston Colloquium for the Philosophy of Science, 1961-1962,* Boston Studies in the Philosophy of Science (ed. by Robert S. Cohen and Marx W. Wartofsky), Volume I. 1973, VIII + 212 pp.
7. A. A. Zinov'ev, *Philosophical Problems of Many-Valued Logic.* 1963, XIV + 155 pp.
8. Georges Gurvitch, *The Spectrum of Social Time.* 1964, XXVI + 152 pp.
9. Paul Lorenzen, *Formal Logic.* 1965, VIII + 123 pp.
10. Robert S. Cohen and Marx W. Wartofsky (eds.), *In Honor of Philipp Frank,* Boston Studies in the Philosophy of Science (ed. by Robert S. Cohen and Marx W. Wartofsky), Volume II. 1965, XXXIV + 475 pp.
11. Evert W. Beth, *Mathematical Thought. An Introduction to the Philosophy of Mathematics.* 1965, XII + 208 pp.
12. Evert W. Beth and Jean Piaget, *Mathematical Epistemology and Psychology.* 1966, XII + 326 pp.
13. Guido Küng, *Ontology and the Logistic Analysis of Language. An Enquiry into the Contemporary Views on Universals.* 1967, XI + 210 pp.
14. Robert S. Cohen and Marx W. Wartofsky (eds.), *Proceedings of the Boston Colloquium for the Philosophy of Science 1964-1966, in Memory of Norwood Russell Hanson,* Boston Studies in the Philosophy of Science (ed. by Robert S. Cohen and Marx W. Wartofsky), Volume III. 1967, XLIX + 489 pp.

15. C. D. Broad, *Induction, Probability, and Causation. Selected Papers.* 1968, XI + 296 pp.
16. Günther Patzig, *Aristotle's Theory of the Syllogism. A Logical-Philosophical Study of Book A of the Prior Analytics.* 1968, XVII + 215 pp.
17. Nicholas Rescher, *Topics in Philosophical Logic.* 1968, XIV + 347 pp.
18. Robert S. Cohen and Marx W. Wartofsky (eds.), *Proceedings of the Boston Colloquium for the Philosophy of Science 1966-1968,* Boston Studies in the Philosophy of Science (ed. by Robert S. Cohen and Marx W. Wartofsky), Volume IV. 1969, VIII + 537 pp.
19. Robert S. Cohen and Marx W. Wartofsky (eds.), *Proceedings of the Boston Colloquium for the Philosophy of Science 1966-1968,* Boston Studies in the Philosophy of Science (ed. by Robert S. Cohen and Marx W. Wartofsky), Volume V. 1969, VIII + 482 pp.
20. J.W. Davis, D. J. Hockney, and W. K. Wilson (eds.), *Philosophical Logic.* 1969, VIII + 277 pp.
21. D. Davidson and J. Hintikka (eds.), *Words and Objections: Essays on the Work of W.V. Quine.* 1969, VIII + 366 pp.
22. Patrick Suppes, *Studies in the Methodology and Foundations of Science. Selected Papers from 1911 to 1969.* 1969. XII + 473 pp.
23. Jaakko Hintikka, *Models for Modalities. Selected Essays.* 1969, IX + 220 pp.
24. Nicholas Rescher *et al.* (eds.), *Essays in Honor of Carl G. Hempel. A Tribute on the Occasion of His Sixty-Fifth Birthday.* 1969, VII + 272 pp.
25. P. V. Tavanec (ed.), *Problems of the Logic of Scientific Knowledge.* 1969, XII + 429 pp.
26. Marshall Swain (ed.), *Induction, Acceptance, and Rational Belief.* 1970, VII + 232 pp.
27. Robert S. Cohen and Raymond J. Seeger (eds.), *Ernst Mach: Physicist and Philosopher,* Boston Studies in the Philosophy of Science (ed. by Robert S. Cohen and Marx W. Wartofsky), Volume VI. 1970, VIII + 295 pp.
28. Jaakko Hintikka and Patrick Suppes, *Information and Inference.* 1970, X + 336 pp.
29. Karel Lambert, *Philosophical Problems in Logic. Some Recent Developments.* 1970, VII + 176 pp.
30. Rolf A. Eberle, *Nominalistic Systems.* 1970, IX + 217 pp.
31. Paul Weingartner and Gerhard Zecha (eds.), *Induction, Physics, and Ethics: Proceedings and Discussions of the 1968 Salzburg Colloquium in the Philosophy of Science.* 1970, X + 382 pp.
32. Evert W. Beth, *Aspects of Modern Logic.* 1970, XI + 176 pp.
33. Risto Hilpinen (ed.), *Deontic Logic: Introductory and Systematic Readings.* 1971, VII + 182 pp.
34. Jean-Louis Krivine, *Introduction to Axiomatic Set Theory.* 1971, VII + 98 pp.
35. Joseph D. Sneed, *The Logical Structure of Mathematical Physics.* 1971, XV + 311 pp.
36. Carl R. Kordig, *The Justification of Scientific Change.* 1971, XIV + 119 pp.
37. Milič Čapek, *Bergson and Modern Physics,* Boston Studies in the Philosophy of Science (ed. by Robert S. Cohen and Marx W. Wartofsky), Volume VII. 1971, XV + 414 pp.

38. Norwood Russell Hanson, *What I Do Not Believe, and Other Essays* (ed. by Stephen Toulmin and Harry Woolf), 1971, XII + 390 pp.
39. Roger C. Buck and Robert S. Cohen (eds.), *PSA 1970. In Memory of Rudolf Carnap*, Boston Studies in the Philosophy of Science (ed. by Robert S. Cohen and Marx W. Wartofsky), Volume VIII. 1971, LXVI + 615 pp. Also available as paperback.
40. Donald Davidson and Gilbert Harman (eds.), *Semantics of Natural Language.* 1972, X + 769 pp. Also available as paperback.
41. Yehoshua Bar-Hillel (ed.), *Pragmatics of Natural Languages.* 1971, VII + 231 pp.
42. Sören Stenlund, *Combinators, λ-Terms and Proof Theory.* 1972, 184 pp.
43. Martin Strauss, *Modern Physics and Its Philosophy. Selected Papers in the Logic, History, and Philosophy of Science.* 1972, X + 297 pp.
44. Mario Bunge, *Method, Model and Matter.* 1973, VII + 196 pp.
45. Mario Bunge, *Philosophy of Physics.* 1973, IX + 248 pp.
46. A. A. Zinov'ev, *Foundations of the Logical Theory of Scientific Knowledge (Complex Logic)*, Boston Studies in the Philosophy of Science (ed. by Robert S. Cohen and Marx W. Wartofsky), Volume IX. Revised and enlarged English edition with an appendix, by G. A. Smirnov, E. A. Sidorenka, A. M. Fedina, and L. A. Bobrova. 1973, XXII + 301 pp. Also available as paperback.
47. Ladislav Tondl, *Scientific Procedures*, Boston Studies in the Philosophy of Science (ed. by Robert S. Cohen and Marx W. Wartofsky), Volume X. 1973, XII + 268 pp. Also available as paperback.
48. Norwood Russell Hanson, *Constellations and Conjectures* (ed. by Willard C. Humphreys, Jr.). 1973, X + 282 pp.
49. K. J. J. Hintikka, J. M. E. Moravcsik, and P. Suppes (eds.), *Approaches to Natural Language. Proceedings of the 1970 Stanford Workshop on Grammar and Semantics.* 1973, VIII + 526 pp. Also available as paperback.
50. Mario Bunge (ed.), *Exact Philosophy – Problems, Tools, and Goals.* 1973, X + 214 pp.
51. Radu J. Bogdan and Ilkka Niiniluoto (eds.), *Logic, Language, and Probability. A Selection of Papers Contributed to Sections IV, VI, and XI of the Fourth International Congress for Logic, Methodology, and Philosophy of Science, Bucharest, September 1971.* 1973, X + 323 pp.
52. Glenn Pearce and Patrick Maynard (eds.), *Conceptual Chance.* 1973, XII + 282 pp.
53. Ilkka Niiniluoto and Raimo Tuomela, *Theoretical Concepts and Hypothetico-Inductive Inference.* 1973, VII + 264 pp.
54. Roland Fraïssé, *Course of Mathematical Logic* – Volume 1: *Relation and Logical Formula.* 1973, XVI + 186 pp. Also available as paperback.
55. Adolf Grünbaum, *Philosophical Problems of Space and Time.* Second, enlarged edition, Boston Studies in the Philosophy of Science (ed. by Robert S. Cohen and Marx W. Wartofsky), Volume XII. 1973, XXIII + 884 pp. Also available as paperback.
56. Patrick Suppes (ed.), *Space, Time, and Geometry.* 1973, XI + 424 pp.
57. Hans Kelsen, *Essays in Legal and Moral Philosophy*, selected and introduced by Ota Weinberger. 1973, XXVIII + 300 pp.
58. R. J. Seeger and Robert S. Cohen (eds.), *Philosophical Foundations of Science. Proceedings of an AAAS Program, 1969*, Boston Studies in the Philosophy of

Science (ed. by Robert S. Cohen and Marx W. Wartofsky), Volume XI. 1974, X + 545 pp. Also available as paperback.

59. Robert S. Cohen and Marx W. Wartofsky (eds.), *Logical and Epistemological Studies in Contemporary Physics*, Boston Studies in the Philosophy of Science (ed. by Robert S. Cohen and Marx W. Wartofsky), Volume XIII. 1973, VIII + 462 pp. Also available as paperback.

60. Robert S. Cohen and Marx W. Wartofsky (eds.), *Methodological and Historical Essays in the Natural and Social Sciences. Proceedings of the Boston Colloquium for the Philosophy of Science, 1969-1972*, Boston Studies in the Philosophy of Science (ed. by Robert S. Cohen and Marx W. Wartofsky), Volume XIV. 1974, VIII + 405 pp. Also available as paperback.

61. Robert S. Cohen, J. J. Stachel and Marx W. Wartofsky (eds.), *For Dirk Struik. Scientific, Historical and Political Essays in Honor of Dirk J. Struik*, Boston Studies in the Philosophy of Science (ed. by Robert S. Cohen and Marx W. Wartofsky), Volume XV. 1974, XXVII + 652 pp. Also available as paperback.

62. Kazimierz Ajdukiewicz, *Pragmatic Logic*, transl. from the Polish by Olgierd Wojtasiewicz. 1974, XV + 460 pp.

63. Sören Stenlund (ed.), *Logical Theory and Semantic Analysis. Essays Dedicated to Stig Kanger on His Fiftieth Birthday*. 1974, V + 217 pp.

64. Kenneth F. Schaffner and Robert S. Cohen (eds.), *Proceedings of the 1972 Biennial Meeting, Philosophy of Science Association*, Boston Studies in the Philosophy of Science (ed. by Robert S. Cohen and Marx W. Wartofsky), Volume XX. 1974, IX + 444 pp. Also available as paperback.

65. Henry E. Kyburg, Jr., *The Logical Foundations of Statistical Inference*. 1974, IX + 421 pp.

66. Marjorie Grene, *The Understanding of Nature: Essays in the Philosophy of Biology*, Boston Studies in the Philosophy of Science (ed. by Robert S. Cohen and Marx W. Wartofsky), Volume XXIII. 1974, XII + 360 pp. Also available as paperback.

67. Jan M. Broekman, *Structuralism: Moscow, Prague, Paris*. 1974, IX + 117 pp.

68. Norman Geschwind, *Selected Papers on Language and the Brain*, Boston Studies in the Philosophy of Science (ed. by Robert S. Cohen and Marx W. Wartofsky), Volume XVI. 1974, XII + 549 pp. Also available as paperback.

69. Roland Fraïssé, *Course of Mathematical Logic* – Volume 2: *Model Theory*. 1974, XIX + 192 pp.

70. Andrzej Grzegorczyk, *An Outline of Mathematical Logic. Fundamental Results and Notions Explained with All Details*. 1974, X + 596 pp.

71. Franz von Kutschera, *Philosophy of Language*. 1975, VII + 305 pp.

72. Juha Manninen and Raimo Tuomela (eds.), *Essays on Explanation and Understanding. Studies in the Foundations of Humanities and Social Sciences*. 1976, VII + 440 pp.

73. Jaakko Hintikka (ed.), *Rudolf Carnap, Logical Empiricist. Materials and Perspectives*. 1975, LXVIII + 400 pp.

74. Milič Čapek (ed.), *The Concepts of Space and Time. Their Structure and Their Development*, Boston Studies in the Philosophy of Science (ed. by Robert S. Cohen and Marx W. Wartofsky), Volume XXII. 1976, LVI + 570 pp. Also available as paperback.

75. Jaakko Hintikka and Unto Remes, *The Method of Analysis. Its Geometrical Origin and Its General Significance*, Boston Studies in the Philosophy of Science (ed. by Robert S. Cohen and Marx W. Wartofsky), Volume XXV. 1974, XVIII + 144 pp. Also available as paperback.
76. John Emery Murdoch and Edith Dudley Sylla, *The Cultural Context of Medieval Learning. Proceedings of the First International Colloquium on Philosophy, Science, and Theology in the Middle Ages – September 1973*, Boston Studies in the Philosophy of Science (ed. by Robert S. Cohen and Marx W. Wartofsky), Volume XXVI. 1975, X + 566 pp. Also available as paperback.
77. Stefan Amsterdamski, *Between Experience and Metaphysics. Philosophical Problems of the Evolution of Science*, Boston Studies in the Philosophy of Science (ed. by Robert S. Cohen and Marx W. Wartofsky), Volume XXXV. 1975, XVIII + 193 pp. Also available as paperback.
78. Patrick Suppes (ed.), *Logic and Probability in Quantum Mechanics*. 1976, XV + 541 pp.
79. Hermann von Helmholtz: *Epistemological Writings. The Paul Hertz/Moritz Schlick Centenary Edition of 1921 with Notes and Commentary by the Editors*. (Newly translated by Malcolm F. Lowe. Edited with an Introduction and Bibliography, by Robert S. Cohen and Yehuda Elkana), Boston Studies in the Philosophy of Science (ed. by Robert S. Cohen and Marx W. Wartofsky), Volume XXXVII. 1977. XXXVIII+204 pp. Also available as paperback.
80. Joseph Agassi, *Science in Flux*, Boston Studies in the Philosophy of Science (ed. by Robert S. Cohen and Marx W. Wartofsky), Volume XXVIII. 1975, XXVI + 553 pp. Also available as paperback.
81. Sandra G. Harding (ed.), *Can Theories Be Refuted? Essays on the Duhem-Quine Thesis*. 1976, XXI + 318 pp. Also available as paperback.
82. Stefan Nowak, *Methodology of Sociological Research: General Problems*. 1977, XVIII + 504 pp.
83. Jean Piaget, Jean-Blaise Grize, Alina Szeminska, and Vinh Bang, *Epistemology and Psychology of Functions*, Studies in Genetic Epistemology, Volume XXIII. 1977, XIV+205 pp.
84. Marjorie Grene and Everett Mendelsohn (eds.), *Topics in the Philosophy of Biology*, Boston Studies in the Philosophy of Science (ed. by Robert S. Cohen and Marx W. Wartofsky), Volume XXVII. 1976, XIII + 454 pp. Also available as paperback.
85. E. Fischbein, *The Intuitive Sources of Probabilistic Thinking in Children*. 1975, XIII + 204 pp.
86. Ernest W. Adams, *The Logic of Conditionals. An Application of Probability to Deductive Logic*. 1975, XIII + 156 pp.
87. Marian Przełęcki and Ryszard Wójcicki (eds.), *Twenty-Five Years of Logical Methodology in Poland*. 1977, VIII + 803 pp.
88. J. Topolski, *The Methodology of History*. 1976, X + 673 pp.
89. A. Kasher (ed.), *Language in Focus: Foundations, Methods and Systems. Essays Dedicated to Yehoshua Bar-Hillel*, Boston Studies in the Philosophy of Science (ed. by Robert S. Cohen and Marx W. Wartofsky), Volume XLIII. 1976, XXVIII + 679 pp. Also available as paperback.
90. Jaakko Hintikka, *The Intentions of Intentionality and Other New Models for Modalities*. 1975, XVIII + 262 pp. Also available as paperback.

91. Wolfgang Stegmüller, *Collected Papers on Epistemology, Philosophy of Science and History of Philosophy*, 2 Volumes, 1977, XXVII + 525 pp.
92. Dov M. Gabbay, *Investigations in Modal and Tense Logics with Applications to Problems in Philosophy and Linguistics.* 1976, XI + 306 pp.
93. Radu J. Bogdan, *Local Induction.* 1976, XIV + 340 pp.
94. Stefan Nowak, *Understanding and Prediction: Essays in the Methodology of Social and Behavioral Theories.* 1976, XIX + 482 pp.
95. Peter Mittelstaedt, *Philosophical Problems of Modern Physics*, Boston Studies in the Philosophy of Science (ed. by Robert S. Cohen and Marx W. Wartofsky), Volume XVIII. 1976, X + 211 pp. Also available as paperback.
96. Gerald Holton and William Blanpied (eds.), *Science and Its Public: The Changing Relationship*, Boston Studies in the Philosophy of Science (ed. by Robert S. Cohen and Marx W. Wartofsky), Volume XXXIII. 1976, XXV + 289 pp. Also available as paperback.
97. Myles Brand and Douglas Walton (eds.), *Action Theory. Proceedings of the Winnipeg Conference on Human Action, Held at Winnipeg, Manitoba, Canada, 9-11 May 1975.* 1976, VI + 345 pp.
98. Risto Hilpinen, *Knowledge and Rational Belief.* 1978 (forthcoming).
99. R. S. Cohen, P. K. Feyerabend, and M. W. Wartofsky (eds.), *Essays in Memory of Imre Lakatos*, Boston Studies in the Philosophy of Science (ed. by Robert S. Cohen and Marx W. Wartofsky), Volume XXXIX. 1976, XI + 762 pp. Also available as paperback.
100. R. S. Cohen and J. J. Stachel (eds.), *Selected Papers of Léon Rosenfeld*, Boston Studies in the Philosophy of Science (ed. by Robert S. Cohen and Marx W. Wartofsky), Volume XXI. 1977, XXX + 927 pp.
101. R. S. Cohen, C. A. Hooker, A. C. Michalos, and J. W. van Evra (eds.), *PSA 1974: Proceedings of the 1974 Biennial Meeting of the Philosophy of Science Association*, Boston Studies in the Philosophy of Science (ed. by Robert S. Cohen and Marx W. Wartofsky), Volume XXXII. 1976, XIII + 734 pp. Also available as paperback.
102. Yehuda Fried and Joseph Agassi, *Paranoia: A Study in Diagnosis*, Boston Studies in the Philosophy of Science (ed. by Robert S. Cohen and Marx W. Wartofsky), Volume L. 1976, XV + 212 pp. Also available as paperback.
103. Marian Przełęcki, Klemens Szaniawski, and Ryszard Wójcicki (eds.), *Formal Methods in the Methodology of Empirical Sciences.* 1976, 455 pp.
104. John M. Vickers, *Belief and Probability.* 1976, VIII + 202 pp.
105. Kurt H. Wolff, *Surrender and Catch: Experience and Inquiry Today*, Boston Studies in the Philosophy of Science (ed. by Robert S. Cohen and Marx W. Wartofsky), Volume LI. 1976, XII + 410 pp. Also available as paperback.
106. Karel Kosík, *Dialectics of the Concrete*, Boston Studies in the Philosophy of Science (ed. by Robert S. Cohen and Marx W. Wartofsky), Volume LII. 1976, VIII + 158 pp. Also available as paperback.
107. Nelson Goodman, *The Structure of Appearance*, Boston Studies in the Philosophy of Science (ed. by Robert S. Cohen and Marx W. Wartofsky), Volume LIII. 1977, L + 285 pp.
108. Jerzy Giedymin (ed.), *Kazimierz Ajdukiewicz: The Scientific World-Perspective and Other Essays, 1931 - 1963.* 1978, LIII + 378 pp.

109. Robert L. Causey, *Unity of Science.* 1977, VIII+185 pp.
110. Richard E. Grandy, *Advanced Logic for Applications.* 1977, XIV + 168 pp.
111. Robert P. McArthur, *Tense Logic.* 1976, VII + 84 pp.
112. Lars Lindahl, *Position and Change: A Study in Law and Logic.* 1977, IX + 299 pp.
113. Raimo Tuomela, *Dispositions.* 1978, X + 450 pp.
114. Herbert A. Simon, *Models of Discovery and Other Topics in the Methods of Science,* Boston Studies in the Philosophy of Science (ed. by Robert S. Cohen and Marx W. Wartofsky), Volume LIV. 1977, XX + 456 pp. Also available as paperback.
115. Roger D. Rosenkrantz, *Inference, Method and Decision.* 1977, XVI + 262 pp. Also available as paperback.
116. Raimo Tuomela, *Human Action and Its Explanation. A Study on the Philosophical Foundations of Psychology.* 1977, XII + 426 pp.
117. Morris Lazerowitz, *The Language of Philosophy, Freud and Wittgenstein,* Boston Studies in the Philosophy of Science (ed. by Robert S. Cohen and Marx W. Wartofsky), Volume LV. 1977, XVI + 209 pp.
118. Tran Duc Thao, *Origins of Language and Consciousness,* Boston Studies in the Philosophy of Science (ed. by Robert S. Cohen and Marx. W. Wartofsky), Volume LVI. 1977 (forthcoming).
119. Jerzy Pelc, *Semiotics in Poland, 1894 - 1969.* 1977, XXVI + 504 pp.
120. Ingmar Pörn, *Action Theory and Social Science. Some Formal Models.* 1977, X + 129 pp.
121. Joseph Margolis, *Persons and Minds, The Prospects of Nonreductive Materialism,* Boston Studies in the Philosophy of Science (ed. by Robert S. Cohen and Marx W. Wartofsky), Volume LVII. 1977, XIV + 282 pp. Also available as paperback.

# SYNTHESE HISTORICAL LIBRARY

Texts and Studies
in the History of Logic and Philosophy

*Editors:*

N. KRETZMANN (Cornell University)
G. NUCHELMANS (University of Leyden)
L. M. DE RIJK (University of Leyden)

1. M. T. Beonio-Brocchieri Fumagalli, *The Logic of Abelard*. Translated from the Italian. 1969, IX + 101 pp.
2. Gottfried Wilhelm Leibniz, *Philosophical Papers and Letters*. A selection translated and edited, with an introduction, by Leroy E. Loemker. 1969, XII + 736 pp.
3. Ernst Mally, *Logische Schriften*, ed. by Karl Wolf and Paul Weingartner. 1971, X + 340 pp.
4. Lewis White Beck (ed.), *Proceedings of the Third International Kant Congress.* 1972, XI + 718 pp.
5. Bernard Bolzano, *Theory of Science*, ed. by Jan Berg. 1973, XV + 398 pp.
6. J. M. E. Moravcsik (ed.), *Patterns in Plato's Thought. Papers Arising Out of the 1971 West Coast Greek Philosophy Conference.* 1973, VIII + 212 pp.
7. Nabil Shehaby, *The Propositional Logic of Avicenna: A Translation from al-Shifā: al-Qiyās*, with Introduction, Commentary and Glossary. 1973, XIII + 296 pp.
8. Desmond Paul Henry, *Commentary on De Grammatico: The Historical-Logical Dimensions of a Dialogue of St. Anselm's.* 1974, IX + 345 pp.
9. John Corcoran, *Ancient Logic and Its Modern Interpretations.* 1974, X + 208 pp.
10. E. M. Barth, *The Logic of the Articles in Traditional Philosophy.* 1974, XXVII + 533 pp.
11. Jaakko Hintikka, *Knowledge and the Known. Historical Perspectives in Epistemology.* 1974, XII + 243 pp.
12. E. J. Ashworth, *Language and Logic in the Post-Medieval Period.* 1974, XIII + 304 pp.
13. Aristotle, *The Nicomachean Ethics.* Translated with Commentaries and Glossary by Hypocrates G. Apostle. 1975, XXI + 372 pp.
14. R. M. Dancy, *Sense and Contradiction: A Study in Aristotle.* 1975, XII + 184 pp.
15. Wilbur Richard Knorr, *The Evolution of the Euclidean Elements. A Study of the Theory of Incommensurable Magnitudes and Its Significance for Early Greek Geometry.* 1975, IX + 374 pp.
16. Augustine, *De Dialectica.* Translated with Introduction and Notes by B. Darrell Jackson. 1975, XI + 151 pp.